Lecture Notes in Computer Science

Commenced Publication in 1973
Founding and Former Series Editors:
Gerhard Goos, Juris Hartmanis, and Jan van Leeuwen

Zdenek Becvar   Robert Bestak
Lukas Kencl (Eds.)

NETWORKING 2012
Workshops

International IFIP TC 6 Workshops
ETICS, HetsNets, and CompNets
Held at NETWORKING 2012
Prague, Czech Republic, May 25, 2012
Proceedings

 Springer

Volume Editors

Zdenek Becvar
Robert Bestak
Lukas Kencl
Czech Technical University in Prague
Department of Telecommunication Engineering
Technicka 2, 166 27 Prague 6, Czech Republic
E-mail: {zdenek.becvar, robert.bestak, lukas.kencl}@fel.cvut.cz

ISSN 0302-9743   e-ISSN 1611-3349
ISBN 978-3-642-30038-7   e-ISBN 978-3-642-30039-4
DOI 10.1007/978-3-642-30039-4
Springer Heidelberg Dordrecht London New York

Library of Congress Control Number: 2012936975

CR Subject Classification (1998): C.2, H.4, D.2, K.6.5, D.4.6, H.3, I.6

LNCS Sublibrary: SL 5 – Computer Communication Networks and Telecommunications

Typesetting: Camera-ready by author, data conversion by Scientific Publishing Services, Chennai, India

Printed on acid-free paper

Springer is part of Springer Science+Business Media (www.springer.com)

Message from the Chairs

The IFIP Networking conference is an international series of events that started in Paris, France, in 2000. The 2012 edition was held at the Czech Technical University in Prague, Czech Republic, in May 2012.

Practically since the beginning, a series of workshops have been co-located with the main conference to discuss timely topics. Three workshops were planned during IFIP Networking 2012: the "Workshop ETICS 2012," the "Workshop HetsNets: Future Heterogeneous Networks 2012," and the "Workshop CompNets: Computing in Networks 2012."

After a strict review process, involving submitted papers plus several invited contributors solicited by the Workshop Chairs, the three workshops took place on May 25, 2012, at the Networking 2012 venue in Prague.

Twenty-one papers were presented and discussed in all three workshops. Fruitful discussions during the workshops were maintained by the Chairs and by the authors as a valuable input for the final writing of the papers. We would like to warmly thank the six workshop Chairs: Helia Pouyllau and Jean-Louis Rougier of ETICS 2012; Emilio Calvanese Strinati and Antonio De Domenico of HetsNets 2012; and Miroslav Voznak and Hakan Kavlak of CompNets 2012. They did excellent work when issuing the call for papers and a very professional management during the review process together with the Technical Program Committees. Thanks to all of them. Last, but not least, we thank all of the attendees for contributing to the success of the workshops.

Thanks to the utmost important topics addressed and the high level of all 21 contributions, this proceedings volume provides a valuable summary of the main research efforts in the fields of network management, quality of services, heterogeneous networks, and analysis or modeling of network.

May 2012

Zdenek Becvar
Robert Bestak
Lukas Kencl

Organization

Executive Committee

General Chairs

Robert Bestak   Czech Technical University in Prague,
Czech Republic

Lukas Kencl   Czech Technical University in Prague,
Czech Republic

Technical Program Chairs

Li Erran Li   Bell Laboratories, Alcatel-Lucent, USA
Joerg Widmer   Institute IMDEA Networks, Spain
Hao Yin   Tsinghua University, China

Publication Chairs

Josep Domenech   Universitat Politècnica de València, Spain
Jan Rudinsky   University of Iceland, Iceland

Finance Chairs

Hana Vyvodova   Czech Technical University in Prague,
Czech Republic

Petr Hofman   Czech Technical University in Prague,
Czech Republic

Publicity Chair

Michal Ficek   Czech Technical University in Prague,
Czech Republic

Workshop Chair

Zdenek Becvar   Czech Technical University in Prague,
Czech Republic

Local Arrangements Chairs

Mylada Balounova   Czech Technical University in Prague,
Czech Republic

Tomas Hegr   Czech Technical University in Prague,
Czech Republic

Webmaster

Marek Nevosad   Czech Technical University in Prague,
Czech Republic

Steering Committee

George Carle   TU Munich, Germany
Marco Conti   IIT-CNR, Pisa, Italy
Pedro Cuenca   Universidad de Castilla-la-Mancha, Spain
Guy Leduc   University of Liège, Belgium
Henning Schulzrinne   Columbia University, USA

ETICS 2012 Chairs

Helia Pouyllau   Alcatel Lucent Bell Labs, France
Jean-Louis Rougier   Telecom ParisTech, France

HetNets 2012 Chairs

Emilio Calvanese Strinati CEA-LETI, France
Antonio De Domenico   CEA-LETI, France

CompNets 2012 Chairs

Miroslav Voznak   Technical University of Ostrava,
                              Czech Republic
Hakan Kavlak   Ericsson Tokyo, Japan

Supporting and Sponsoring Organizations

Faculty of Electrical Engineering, Czech Technical University in Prague
IFIP TC 6
Microsoft Research

ETICS 2012 Technical Program Committee

Johanne Cohen   UVSQ, France
Costas Courcoubetis   AUEB, Greece
Richard Douville   Alcatel Lucent Bell Labs, France
Olivier Dugeon   Orange Labs, France
Antionio Ghezzi   Politecnico di Milano, Italy
Ivan Gojmerac   FTW, Austria
Isabelle Korthals   Deutsche Telekom Laboratories, Germany
Nicolas Le Sauze   Alcatel Lucent Bell Labs, France
Ariel Orda   Technion, Israel
Peter Reichl   FTW, Austria
George Stamoulis   AUEB, Greece
Sandrine Vatton   Telecom Bretagne, France

HetNets 2012 Technical Program Committee

Josep Vidal	UPC, Spain
Sergio Barbarossa	University of Rome, Italy
Jean-Marie Gorce	INRIA, France
Mischa Dohler	CTTC, Spain
Vikram Chandrasekhar	Texas Instruments, USA
David López-Pérez	King's College, UK
D'Errico Raffaele	CEA-LETI, France
Antonio Cappone	University of Milan, Italy
Marios Kountouris	SUPELEC, France
Mehedi Bennis	CWC, University of Oulu, Finland
Cédric Abgrall	CEA-LETI, France
Tony Quek	Institute for Infocomm Research, Singapore
Magnus Olsson	Ericsson, Sweden
Atta Quddus	CCSR University of Surrey, UK
Chiara Petrioli	University of Rome, Italy
Konstantinos Dimou	Ericsson, Sweden
Haibin Zhang	TNO, The Netherlands
Simone Morosi	University of Florence, Italy
Dominique Noguet	CEA-LETI, France
Dario Sabella	Telecom Italia Labs, Italy
Dario Rossi	Telecom Paris Tech, France
Guillaume Vivier	Sequans, France

CompNets 2012 Technical Program Committee

Andrzej Dziech	AGH University of Science and Technology, Krakow, Poland
Kathiravelu Ganeshan	Unitec Institute of Technology, New Zealand
Hakki Gokhan Ilk	Ankara University, Turkey
Dan Komosny	Brno University of Technology, Czech Republic
Victor Rangel Licea	National Autonomous University of Mexico, Mexico
Mikolaj Leszczuk	AGH University of Science and Technology, Krakow, Poland
Zafer Sahinoglu	ME Research Labs, USA
Ibrahim Taner Okumus	Sutcu Imam University, Turkey
Ivan Zelinka	Technical University of Ostrava, Czech Republic

Workshop on Economics and Technologies for Inter-Carrier Services (ETICS 2012)

The rapid evolution of Internet applications is causing major changes in carriers' business and network management. To support such demands (e.g., gaming, tele-conference), telco operators have moved their business strategy to provide more applications and services to both end-users and business customers. This strategy change, strengthened by other contradictions in the Internet business model, has generated more tussles with application providers for network resources or with other Internet service providers. As some of these tensions were publicly disclosed, many discussions arose worldwide around net neutrality –application providers asking regulation authorities to keep networks neutral and telco operators arguing for the need for investment incentives. Moreover, application needs for QoS guarantees have become increasingly urgent, especially in the field of real-time interactive applications and for supporting security impediments.

The main objective of the ETICS 2012 workshop was to bring together economic and technical experts of the Internet to discuss some of the key economic and technical issues to alleviate the aforementioned tussles.

We would like to thank the IFIP Networking 2012 organizers for their help in the organization of this event. We would also like to thank the European Commission for funding the ETICS project and thus the venue and keynote speakers as well. Finally, we also would like to acknowledge the TPC members, the authors and the workshop attendees.

May 2012

Helia Pouyllau
Jean-Louis Rougier

Workshop on Future Heterogeneous Networks (HetsNets 2012)

The workshop on Future Heterogeneous Networks 2012 (HetsNets 2012) was held in Prague, Czech Republic, on May 25, 2012, in conjunction with the 11th International Conferences on Networking 2012 sponsored by the IFIP Technical Committee on Communication Systems (TC6).

Future deployments of heterogeneous cellular networks, which support macro, pico, relay, femto, and atto cells coexisting on the same spectrum in the same geographical area, lead to new technical challenges never faced before. To address these issues, industry and academia are working on the development of new technologies and cellular standards.

The main objective of the workshop is to offer an opportunity to academic and industrial researchers to spread and share results and knowledge in the area in order to make communication networks more efficient than they are today. The workshop consisted of seven papers with contributions from France, Germany, Ireland, Italy, The Netherlands, Pakistan, Portugal, Serbia, and Spain.

We would like to thank Robert Bestak and Lukas Kencl, the General Chairs of Networking 2012, and Zdenek Becvar, the Workshop Chair of Networking 2012, all from Czech Technical University in Prague, for their encouragement and support in organizing this successful workshop. Our gratitude also goes to our TPC members and reviewers for their excellent cooperation.

May 2012

Emilio Calvanese Strinati
Antonio De Domenico

Workshop on Computing in Networks (CompNets 2012)

The aim of the Computing in Networks 2012 workshop is to provide a forum to present new approaches in the area of the analysis, modeling, behavior visualization and network simulations.

Today, a substantial amount of data is surrounding us, generated both by human action and by technology. Such data can be a source of interesting information which is not visible at first glance. Data from these sources can be interpreted as a result of the network behavior (computer or communication network, transport, information, social and bio-inspired network). Very often, these are extensive data, in many cases they indicate issues that require designing new models, performing simulations and predicting network behavior. The complexity and extent of such solutions are the reason why these issues cannot be solved directly and it is necessary to design heuristic schemes and to apply soft computing methods.

The Computing in Networks 2012 workshop focused primarily on the application of new approaches and the optimization of known methods for analyzing extensive network data with the emphasis on demos.

We would like to appreciate the contribution of the authors of all submitted papers and also to thank the Technical Program Committee for their reviews and valuable remarks. Last but not least, we would like to especially thank the IFIP Networking 2012 organizers for the support and help they provided to us.

May 2012

Miroslav Voznak
Hakan Kavlak

Table of Contents

Future Heterogeneous Network

Economics and Technologies for Inter-Carrier Services

Computing in Networks

A Fairness Model for Resource Allocation in Wireless Networks

Huaizhou Shi, R. Venkatesha Prasad, Vijay S. Rao, and I.G.M.M. Niemegeers

Wireless and Mobile Communications, Faculty of EEMCS,
Delft University of Technology, The Netherlands

Abstract. In wireless networks many nodes contend for available resources creating a challenge in resource allocation. With shared resources, fairness in allocation is a serious issue. Fairness metrics have been defined to measure the fairness level of resource decisions in allocations. Therefore, fairness metrics significantly influence network and node performances emphasizing the need for due diligence to fairness metrics. It can be seen that fairness metrics in many of the allocation strategies and algorithms in the literature are not analyzed in depth to show the overall fairness of allocations from different perspectives. Hence, we propose a series of fairness metrics for resource allocation in wireless networks, which evaluate individual, system, short and long term fairness. Our metrics are general enough to be adapted in either multiple or single resource sharing scenarios. Algorithms using these metrics for channel allocation problem in peer-to-peer wireless regional area network is proposed and simulated, and the results show that our metrics in conjunction with smart channel allocation strategies guarantee both the fairness and performances.

Keywords: Fairness metrics, Resource allocation, Wireless networks, Jain's index, P2PWRAN.

1 Introduction

In wireless networks, many resources have to be shared. While in an individual wireless device or node, its computing resources and memory are shared among different applications, at the network level different layers of the device have to operate in synergy with other nodes in order to achieve successful communication and task completion. In a wireless network, channels, time slots, bandwidth, and other resources need to be allocated to different devices in either centralized or distributed way. Hence, resource allocation plays an important role in wireless networking.

Two optimization goals of resource allocation are utility and fairness [6]. Utility reflects the performance of wireless network, and fairness indicates the balance in resource sharing by different wireless devices. Much work can be seen in literature in this regard with emphasis of fairness in resource allocation [7,6]. Though a detailed discussion on fairness issues for resource allocation in wireless networks covering most of the fairness issues is in [2,7], there is still a lack of general fairness model. Some issues still need more considerations, for example,

Z. Becvar et al. (Eds.): NETWORKING 2012 Workshops, LNCS 7291, pp. 1–9, 2012.
© IFIP International Federation for Information Processing 2012

not treating fairness as an independent issue [6], confusing it with resource allocation itself or utility [9], and only considering absolute fairness during allocation in which weights are not adopted or every individual is assigned with the same weight [2,7,3].

In this paper, we discuss related work on fairness metrics in wireless networks in Section 2. We propose a new fairness model for resource allocation that contains a series of fairness metrics in Section 3. These metrics provide an entire view of fairness with multiple fairness metrics. The simulations in Section 4 confirm that our metrics can be employed in resource allocation in wireless networks easily. Conclusions are provided in Section 5.

2   Related Work on Fairness Issues

Some commonly adopted fairness metrics in wireless networks in the literature are Jain's index, max-min and proportional fairness [2,7,3]. Jain's index does not consider the individual fairness. Another shortcoming is that it cannot be adopted in multiple resource allocation scenarios. Moreover, weighted strategies are not considered in it, and all individuals are treated similarly during allocations, which is hardly true in reality. Max-min indicates a decision $0, 1$ whether the allocation is fair or not (1 implying fair allocation). Thus, max-min fairness cannot measure the level of unfairness. Also, it cannot distinguish between system and individuals. Similar to Jain's index, it is an absolute fairness metric without weighted strategies and individual fairness. Modified max-min fairness are proposed (for example in [1]) to add weights, but they cannot measure the individual fairness. Proportional fairness, however, cannot measure how unfair the system is, and no weighted strategies are considered originally. An advanced proportional fairness metric $(p, \alpha) - proportional$ fairness was proposed in [5]. Weights of different resources for an individual are added in this model. However, it cannot measure both individual and system fairness. Moreover, assigning the weights to individuals was not described in detail.

Considering the shortcomings in these existing fairness metrics, we try to propose a fairness model which includes a series of fairness metrics for resource allocation that can measure the fairness of a resource allocation decision in the aspects of system, individual, short, and long term fairness.

3   Fairness Model

We assume that there are m $(m \geqslant 1)$ different resources, and the capacity of resource k is $C_k^{(t)}$. n individuals are in the system. The allocation decision at time t is $\mathbb{X}^{(t)} = (X_1^{(t)}, ..., X_n^{(t)})$, in which $X_k^{(t)}$ is the allocation of resource k and $X_k^{(t)} = (x_{k1}^{(t)}, ..., x_{kn}^{(t)})$. $x_{ki}^{(t)}$ is the amount of resource k that is allocated to individual i. Furthermore, $F_k^{(t)}(\mathbb{X}^{(t)})$ is the system fairness metric on resource k and $f_{ki}^{(t)}(\mathbb{X}^{(t)})$ is the fairness metric for an individual i for resource k. $f_i^{(t)}$ is the fairness of individual i on all resources. When $m = 1$, it is single resource allocation and we just drop the resource subscript, for example, $C^{(t)}$ is the resource capacity, $\mathbb{X}^{(t)} = (x_1^{(t)}, ..., x_n^{(t)})$ is the allocation decision for an individual i. We consider two elements during the weight assignment, which are request ratios and historical allocation information. For individuals with more requests and less historical allocation than others, heavier weights should be assigned to them, giving them higher chance to obtain more resources. Let $\mathbb{R}^{(t)} = \{R_1^{(t)}, ..., R_m^{(t)}\}$ stand for request at time t, in which $R_k^{(t)}$ is the allocation request for resource k. $R_k^{(t)} = (r_{k1}^{(t)}, ..., r_{kn}^{(t)})$ and $r_{ki}^{(t)}$ is the request of resource k from individual i at time t. We assume that $x_{ki}^{(t)} \leqslant r_{ki}^{(t)} \leqslant C_k$. We also assume the weights are $\mathbb{W}^{(t)} = \{W_1^{(t)}, ..., W_m^{(t)}\}$ and $W_k^{(t)} = (w_{k1}^{(t)}, ..., w_{km}^{(t)})$, in which $W_k^{(t)}$ is the weights of resource k, and $w_{ki}^{(t)}$ is the weight of individual i for resource k at time t. We assign the weight $w_{ki}^{(t)}$ as,

$$
w_{ki}^{(t)} = \begin{cases} 0 & \text{If } r_{ki}^{(t)} = 0; \\ e^{\left(1 - \frac{\sum_i r_{ki}^{(t)}}{n r_{ki}^{(t)}}\right)} & \text{Else if } \sum_{t,i} x_{ki}(t) = 0; \\ e^{\left(1 - K_r \frac{\sum_i r_{ki}^{(t)}}{n r_{ki}^{(t)}} - K_h \frac{n \sum_{t, r_{ki}^{(t)} \neq 0} \frac{x_{ki}^{(t)}}{r_{ki}^{(t)}}}{\sum_{t,i,r_{ki}^{(t)} \neq 0} \frac{x_{ki}^{(t)}}{r_{ki}^{(t)}}}\right)} & \text{Otherwise.} \end{cases} \tag{1}
$$

Where $w_{ki}^{(t)}$ indicates the importance and priority for a resource k set by an individual i at time t by considering the request and historical allocation. K_r and K_h ($K_r \leq 0$ and $K_h \leq 0$) are the factors for the current request and historical allocation, and $Kr + k_h = 1$. Therefore, $0 \leqslant w_{ki}^{(t)} \leqslant e$. Based on the weights and request, we define "fair ratio" $e_{ki}^{(t)}$ in Eq. (2), which indicates the satisfaction level of an individual to the current allocation.

$$
e_{ki}^{(t)} = \begin{cases} e^2 & \text{If } r_{ki}^{(t)} w_{ki}^{(t)} = 0; \\ \frac{e^{\left(\frac{x_{ki}^{(t)}}{r_{ki}^{(t)}}\right)}}{w_{ki}^{(t)}} & \text{Otherwise.} \end{cases} \tag{2}
$$

All our following short and long term and individual and system fairness metrics in this paper are based on fair ratios ($e_{ki}^{(t)}$). Incase where there is no request or weight is zero, an individual is satisfied with ($e_{ki}^{(t)} = e^2$), otherwise, the satisfaction depends on the ratio of its allocation and weight.

The short term system fairness metric for resource k at allocation t (can also be a discrete allocation number) is given by Eq. (3) where $0 \leqslant F_k(X_k^{(t)}) \leqslant 1$. It is a variant of Jain's index with the weights and can measure fairness even in multiple resources scenarios.

$$
F_k^{(t)}(X_k^{(t)}) = \begin{cases} 0 & \text{If } \sum_i e_{ki}^{(t)} = 0; \\ 1 & \text{Else if } \sum_i w_{ki}^{(t)} r_{ki}^{(t)} = 0; \\ \frac{\left(\sum_i e_{ki}^{(t)}\right)^2}{n \sum_i \left(e_{ki}^{(t)}\right)^2} & \text{Otherwise.} \end{cases} \tag{3}
$$

The short term individual fairness of individual i on resource k is shown in Eq. (4), which measures the current fairness of one individual over a certain resource, and $0 \leqslant f_{ki}^{(t)} \leqslant 1$, but with the historical allocation information.

$$
f_{ki}^{(t)} = \begin{cases} 0 & \text{If } e_{ki}^{(t)} = 0; \\ e^{-\frac{\sum_i e_{ki}^{(t)}}{n\, e_{ki}^{(t)}}} & \text{Otherwise.} \end{cases}
\tag{4}
$$

The individual fairness of i for all resources is described in Eq. (5), which is the fairness metric for an individual in the whole resource allocation with multiple resources.

$$
f_i^{(t)} = \begin{cases} 0 & \text{If } \sum_k e_{ki}^{(t)} = 0; \\ e^{-\frac{\sum_{k,i} e_{ki}^{(t)}}{n \sum_k e_{ki}^{(t)}}} & \text{Otherwise.} \end{cases}
\tag{5}
$$

Similar to short term fairness metrics, system and individual long term metrics are proposed in Eq. (6), (7), and (8), in which the fairness ratios $e_{ki}^{(t)}$ are considered in a long time period instead of a single time slot.

$$
F_k(X_k) = \begin{cases} 0 & \text{If } \sum_{t,i} e_{ki}^{(t)} = 0; \\ 1 & \text{Else if } \sum_{t,i} r_{ki}^{(t)} = 0; \\ \dfrac{\left(\sum_{t,i} e_{ki}^{(t)}\right)^2}{n \sum_i \left(\sum_t e_{ki}^{(t)}\right)^2} & \text{Otherwise.} \end{cases}
\tag{6}
$$

$$
f_{ki} = \begin{cases} 0 & \text{If } \sum_t e_{ki}^{(t)} = 0; \\ e^{-\frac{\sum_{t,i} e_{ki}^{(t)}}{n \sum_t e_{ki}^{(t)}}} & \text{Otherwise.} \end{cases}
\tag{7}
$$

$$
f_i = \begin{cases} 0 & \text{If } \sum_{t,k} e_{ki}^{(t)} = 0; \\ e^{-\frac{\sum_{t,k,i} e_{ki}^{(t)}}{n \sum_{t,k} e_{ki}^{(t)}}} & \text{Otherwise.} \end{cases}
\tag{8}
$$

Our model is proposed keeping in mind multiple resource allocation scenarios. As a special case for single resource allocation, we just need to set $m = 1$, and all metrics and the mechanisms stay the same. For distributed scenarios, allocation decisions are made by individuals instead of a centralized allocation manager. Therefore, the metrics only use the local information they collect, and localized fairness and utility can be achieved. The following advantages can be seen in our fairness metric for resource allocation:

- We adopt a weighted strategy and define it by considering the request and historical allocations, which assign priority to individuals due to their requests and historical allocation information.
- Both system and individual fairness can be measured by our metrics, which describe fairness on different aspects.

– The fair ratio $e_{ki}^{(t)}$ considers the satisfaction of individuals in an allocation, which gives a more precise description of fairness feature than absolute fairness [2, 7, 3].
– Short term and long term fairness can be measured by our metrics, which may be used in different applications.
– Our fairness model is proposed for multiple resource fair allocation in wireless networks; however, it can also be adapted to other diverse scenarios with slight modification.

4 Simulations and Results

4.1 An Example Scenario

Peer to peer WRAN (P2PWRAN) was proposed in [8]. P2PWRANs support peer to peer communication in a cell thereby increasing the network capacity by many folds. At the same time, a problem in P2PWRANs is how to allocate the channels efficiently and fairly. Channel allocation has two parameters - (i) total number of available channels, and (ii) interference amongst inter-CPE communication[1]. It is not easy to achieve a fair resource allocation at all times. Therefore, we test our definition of fairness metric model in the channel allocation scenario in a P2PWRAN cell.

Two types of communication can be seen in P2PWRAN communication in a cell, which are communication between CPEs and between CPE to BS. The allocation of channels for CPEs can be treated as a multiple resource allocation case. The first type is the channel allocation for CPE to CPE communication and the other type is the channel allocation for CPE to BS (and BS to CPE) communication. Channel allocation in standard WRAN is queue based [4]. Hence, we first queue the channel requests from CPEs, then allocate channel sequentially to the request if no interference is expected to occur. Three different mechanisms for queuing channel requests were adopted in the channel allocation for P2PWRAN in our simulations as follows.

– First one (*M1*) is the absolute fair queuing mechanism, in which the requests are queued considering the channel allocation history of the nodes. CPEs which have received lesser channels have more chance of getting a channel in the current time slot. *M1* treats every individual identically, and tries to reach absolute fairness amongst all individuals during channel allocation.
– The second mechanism (*M2*) is based on our long term individual fairness metric. The long term individual fairness is employed as the basis for queuing. With (*M2*), CPEs which are not treated fairly in the past would have more opportunities to obtain channels.
– The third mechanism (*M3*) queues the requests due to the possibilities of the interference that the requests may cause. Since, if the same channel is allocated to two request (r_i, r_j) both of which cause less interference than

---

[1] For a more detailed discussion of channel allocation in P2PWRAN we refer to [8].

two other requests (r'_i, r'_j), then it is better to allocate the channel to r_i and r_j. Generally, requests with shorter transmission distances cause lesser interference than requests with longer distances. Therefore, the transmission distances of the requests are used to queue channel requests, and requests with short transmission distances are allocated first. The purpose of *M3* is to reuse channels efficiently.

4.2   Results

The network performance on channel reuse and node throughput can be seen in Fig. 1. In Fig. 1(a) and 1(b), the queue mechanism based on our fairness metrics lead to the higher channel reuse times for every channel and most of the nodes obtain higher throughput than the absolute fair mechanism and distance based mechanisms. With our metric in *M2*, node requests are queued considering individual long term fairness and the ratios of node requests and historical allocation decisions, and it leads to the best performance amongst the three mechanisms. We can also see in the figures that fair mechanisms performs better than the distance based mechanisms (*M3*).

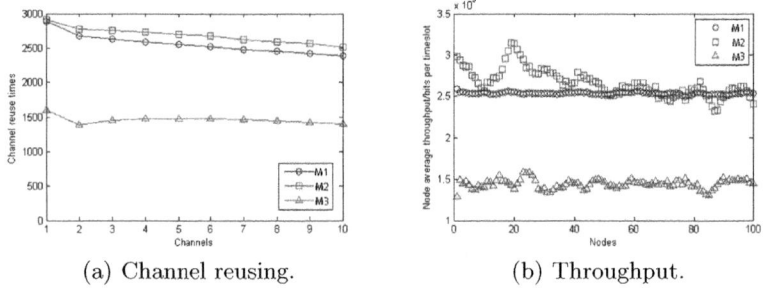

(a) Channel reusing.   (b) Throughput.

Fig. 1. Performances

The individual fairness of channel allocation is shown in Fig. 2 with both short term (Fig. 2(b) and Fig. 2(d)) and long term fairness (Fig. 2(a) and Fig. 2(c)). For CPE to CPE communication, the mechanism based on our metrics (*M2*) results in fairer allocation than absolute fair mechanism *M1* and distance based mechanism *M3* in both short term and long term fairness. The individual long term fairness is adopted to queue the request, therefore, it results in fairer individual allocation in CPE to CPE communication. However, the short term fairness is still much lower than the long term fairness, because the channels are very rare compared to amount of requests. For CPE to BS allocation, the amount of requests (on average more than 20 requests in a time slot) is much higher than the allocation rate (once in a time slot). Hence, both long term and short term CPE to BS fairness are very low, as shown in Fig. 2(d) and Fig. 2(c).

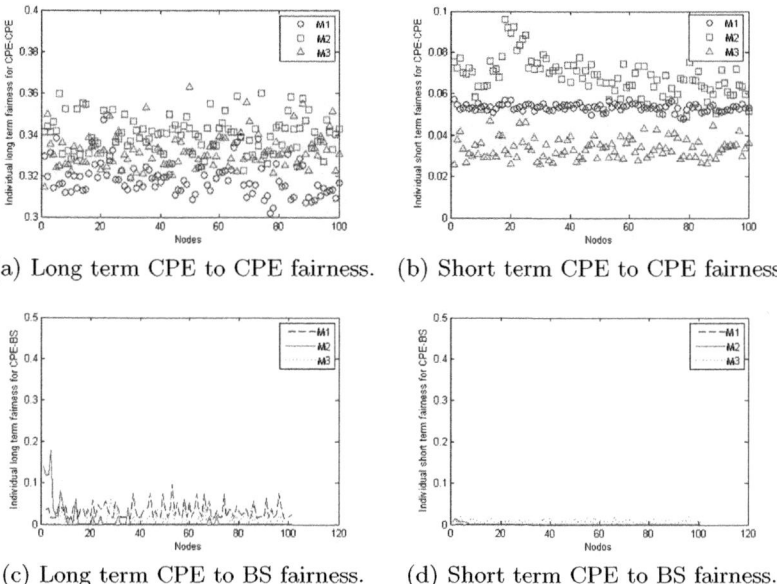

(a) Long term CPE to CPE fairness.   (b) Short term CPE to CPE fairness.

(c) Long term CPE to BS fairness.   (d) Short term CPE to BS fairness.

Fig. 2. Individual fairness

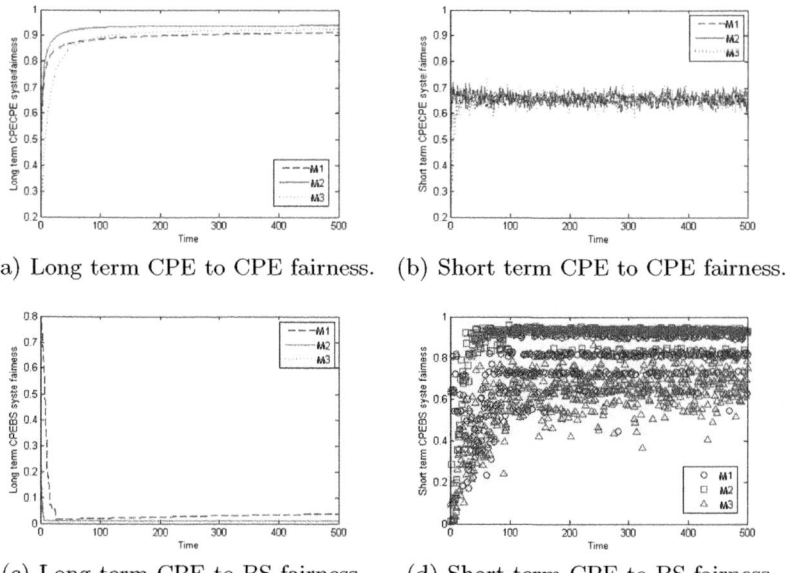

(a) Long term CPE to CPE fairness.   (b) Short term CPE to CPE fairness.

(c) Long term CPE to BS fairness.   (d) Short term CPE to BS fairness.

Fig. 3. System fairness

Results of system fairness are shown in Fig. 3 including both short term and long term system fairness. For CPE to CPE communication, long term fairness is adopted to queue the requests, therefore, mechanism ($M2$) leads to fairer allocation than $M1$ and $M3$ as shown in Fig. 3(a). The short term system fairness for CPE to CPE with $M1$, $M2$ and $M3$ are similar to each other as shown in Fig. 3(b). The long term system fairness for CPE to BS (Fig. 3(c)) are very low because only one node can obtain a channel in one time slot, and the fairness for the rest is assigned to be 0. The short term fairness for CPE to BS is increased by $M2$ (Fig. 3(d)) because in many time slots there is no CPE to BS request, and the short term system fairness is assigned as 1.

5   Conclusions and Further Work

Concerning all the resources and their allocation, fairness issue plays an important role in wireless networks. However, the metrics nowadays has many shortages. Therefore, a series of fairness metrics are proposed in this paper for resource allocation in wireless networks, which can measure the fairness of allocation from the point of view of individuals and system, further, also from the point view of short term and long term. We applied the developed fairness metric for IEEE 802.22 WRAN scenario. We also applied our metric to test the allocation fairness in a futuristic P2P channel usage in WRANs. The simulation results show that the metrics can be employed in wireless resource allocation mechanisms easily. When it is adopted in the channel allocation in P2PWRAN, higher fairness and performance can be achieved than the fairness and performance when absolute and distance based mechanisms are used. We are now planning to focus on more number of implementation scenarios (for example in rate allocation, flow and packet scheduling, power control and energy control) to test our metric, and testing its ability of measuring fairness.

References

1. Bertsekas, D.P., Gallager, R., Nemetz, T.: Data networks. Prentice-Hall, Englewood Cliffs (1987)
2. Jain, R., Chiu, D., Hawe, W.: A Quantitative Measure of Fairness and Discrimination for Resource Allocation in Shared Systems, Digital Equipment Corporation. Tech. rep., Technical Report DEC-TR-301 (1984)
3. Kelly, F., Maulloo, A., Tan, D.: Rate control for communication networks: shadow prices, proportional fairness and stability. The Journal of the Operational Research Society 49(3), 237–252 (1998)
4. Ko, G., Franklin, A., You, S., Pak, J., Song, M., Kim, C.: Channel management in IEEE 802.22 wran systems. IEEE Communications Magazine 48(9), 88–94 (2010)
5. Mo, J., Walrand, J.: Fair end-to-end window-based congestion control. IEEE/ACM Transactions on Networking (ToN) 8(5), 556–567 (2000)
6. Montuno, K., Zhacfi, Y.: Fairness of Resource Allocation in Cellular Networks: A Survey. In: Resource Allocation in Next Generation Wireless Networks, pp. 249–266 (2006)

7. Radunovic, B., Le Boudec, J.Y.: A unified framework for max-min and min-max fairness with applications. IEEE/ACM Transactions on Networking 15(5), 1073–1083 (2007)
8. Shi, H., Prasad, R.R.V., Niemegeers, I.G.: An intra-cell peer to peer protocol in ieee 802.22 networks. In: GLOBALCOM (GC 2011) Workshop on Mobile Computing and Emerging Communication Networks (December 2011) (accepted)
9. Zheng, D., Zhang, J.: A two-phase utility maximization framework for wireless medium access control. IEEE Transactions on Wireless Communications 6(12), 4207–4299 (2007)

An Extension and Cooperation Mechanism for Heterogeneous Overlay Networks[*]

Vincenzo Ciancaglini[1,2,**], Luigi Liquori[1],
Giang Ngo Hoang[1,2], and Petar Maksimović[1,3]

[1] Institut National de Recherche en Informatique et Automatique, France
[2] Hanoi University of Science and Technology, Vietnam
[3] Mathematical Institute of the Serbian Academy of Sciences and Arts, Serbia
First.Last@inria.fr

Abstract. In real-world peer-to-peer applications, the scalability of data lookup is heavily affected by network artifacts. A common solution to improve scalability, robustness and security is to increase the local properties of nodes, by clustering them together. This paper presents a framework which allows for the development of distributed applications on top of interconnected overlay network. Here, message routing between overlays is accomplished by using co-located nodes, i.e. nodes belonging to more than one overlay network at the same time. These co-located nodes serve as distributed gateways, enabling the routing of requests across overlays, while keeping overlay maintenance operations local. The protocol has been evaluated via simulations and client deployment, showing that the ability, of reaching the totality of the overlays in a federated configuration can be preserved even with the simplest routing, proving the feasibility of federated overlay configurations.

Keywords: Peer-to-peer, overlay networks, interoperability.

1   Introduction and Related Work

Context. Overlay networks, structured and unstructured alike, have been broadly recognized as one viable solution for the implementation of various distributed applications: Distributed Hash Tables (DHTs), application-level multicast protocols, distributed object lookup services, file systems, etc. Although most of the protocols designed during the recent years have the theoretical properties of scalability, fault tolerance and handling of dynamic topologies, studies such as [10] have shown that, under real networking conditions, overlays are still vulnerable to severe performance degradation once deployed on a larger scale. The general problem on which we focus concerns the viability of deploying networks consisting of multiple smaller structured overlays, rather than one global overlay, which can interact with each other, thus exploiting locality and diverse topologies: being able to reduce the scope of overlay maintenance to a smaller diameter can benefit performance and fault tolerance, while

---

[*] Work partially supported by the Serbian Ministry of Education and Science (projects ON 174026 and III 044006).
[**] Corresponding author.

Z. Becvar et al. (Eds.): NETWORKING 2012 Workshops, LNCS 7291, pp. 10–18, 2012.
ⓒ IFIP International Federation for Information Processing 2012

having a mechanism allowing nodes to reach overlays other than their own facilitates the extensibility and flexibility of their design. The same mechanism could then also be exploited for the cooperation of existing overlay networks.

Related Work. Efficient cooperation between heterogeneous overlays has served as an inspiration to a number of research efforts, which include concepts such as *cooperation via gateway* (in [5, 6]), *cooperation via super-overlay* (in [11, 15]), *cooperation via hierarchy* (in [8, 9, 16]), *merging of overlays* (in [6, 7, 14]), and *hybrid overlay systems* (in [4, 13]). These efforts and concepts have been described and contrasted to our approach in more detail in [17], while here we only mention that we share their common idea of increasing locality in the network, and have the most features in common with the 'cooperation via hierarchy' approach.

Synapse. In [12], we have introduced a protocol, which we have named *Synapse*, capable of interconnecting heterogeneous overlays through utilization of co-located nodes as distributed gateways. This protocol was further explored and improved in [17], resulting in the *Synapse Framework*, which consists of the related Synapse protocol, software architecture, libraries in the OverSim simulator [3], and a peer client written in Java. This framework allows for a flexible implementation of different federated topologies, as well as the application of various routing strategies across these topologies, depending on the application scenario in question. In this workshop paper, we present a first evaluation of the Synapse protocol, accomplished via extensive simulations in OverSim, and deployment on the Grid'5000 platform of one of its possible routing strategies, namely 1-random-walk, on a topology comprising a cluster of heterogeneous structured Chord and Kademlia overlays. From the obtained results, we gain several insights concerning the design of such systems.

Summary. The remainder of the paper is structured as follows: in Section 2 we provide a brief description of the Synapse framework, as well as the routing strategy used for the simulations. Next, in Section 3 we present and discuss the results obtained from simulations performed in the OverSim simulator, and from the deployment of Synapse on the Grid'5000 platform. Finally, in Section 4 we give our conclusions and discuss further work. All of the details concerning the Synapse framework, as well as the full results of simulations and experiments performed are available at [2].

2   The Synapse Framework

In this section, we briefly present the inter-overlay cooperation protocol which is part of the Synapse Framework, and the specific routing strategy implemented in our tests. Due to space restrictions, here we only provide an outline, while for a more articulate system description, we refer the reader to [17].

Synapse Network Topology. The Synapse Framework enables the creation of a network made of multiple overlays, each identified by a unique `netID`, capable of interacting with each other. Each overlay can exploit a different topology, different data indexing schemes, and different maintenance mechanisms.

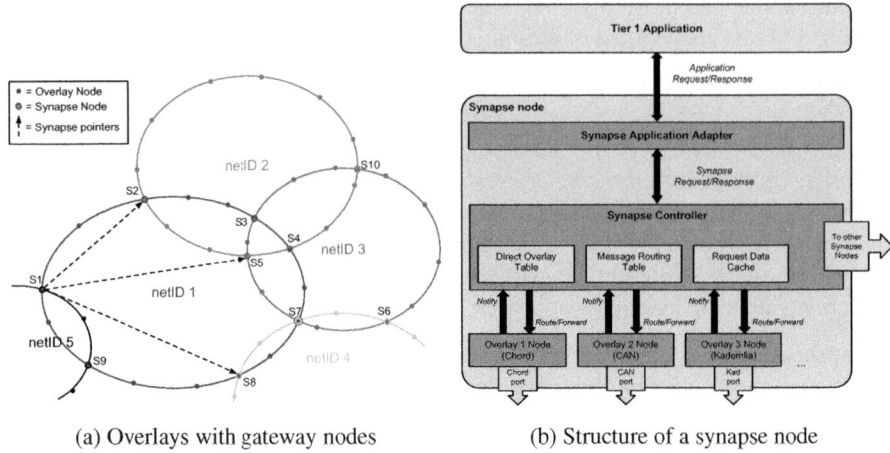

(a) Overlays with gateway nodes   (b) Structure of a synapse node

Fig. 1. Inter-overlay topologies and structure of synapse nodes

Cooperation and message routing between overlays is achieved via co-located nodes which are connected to two or more networks at the same time, and can perform requests on behalf of other nodes. We will refer to these nodes as *gateway nodes*. Also, we will be referring to nodes which are aware of the Synapse protocol as *synapse nodes*. Finally, as the inter-routing of messages requires data other than the overlay-specific one to be exchanged between nodes, there are feasible scenarios in which, due to the interconnection of existing overlays (e.g. the Mainline DHT of BitTorrent, in [1]), there can be nodes which are unable to understand the additional data embedded in the protocol messages. We refer to such nodes as *legacy nodes*.

In the Synapse Framework, synapse nodes and gateway nodes form an unstructured overlay, with each synapse node keeping a neighborhood of gateway nodes pointing to foreign overlays; this is used to exchange inter-routing requests and other information between the synapse nodes and gateway nodes. An example of a possible Synapse topology made up of 5 overlays, interconnected via a number of gateway nodes, is shown in Figure 1a.

Synapse Node Architecture. The architecture of a synapse node is shown in Figure 1b. A node usually maintains different instances of virtual overlay nodes, one for each overlay, and additional data structures to handle the inter-routing, namely a *Direct Overlay Table (DOT)* with pointers to gateway nodes arranged by overlay, a *Message Routing Table* to keep track of ongoing requests, and a *Request Data Cache*. Synapse and gateway nodes share the same architecture, with the only difference being in the number of overlays the node is connected to. This makes it possible to dynamically adjust the connectivity of each overlay, by increasing the number of overlays that nodes connect to. The effect is achieved via *social-based primitives*, i.e. invitation messages issued to gateway node candidates, asking them to increase their connectivity, and join a specific overlay.

Routing in Synapse. Whereas legacy nodes can only perform requests in the overlay they are connected to, synapse nodes are capable of reaching beyond their connected overlays by contacting gateway nodes. We have, therefore, two different routing mechanisms, one consisting simply of using the virtual node instances to issue a request in a connected overlay, and a second one consisting of sending a SYNAPSE_REQUEST message to known gateways. A SYNAPSE_REQUEST message carries various parameters within itself (e.g. *Request ID, TTL, Routing History, Target Overlay List*) which can be used to implement different routings between overlays, depending on the application and the desired QoS. The choice of gateway nodes at each routing step, and the choice of how should the request propagate and which overlays should it target is what constitutes a *Routing Strategy* in Synapse.

As mentioned above, routing strategies are application-dependent: for example, an application can decide to issue inter-routing requests following a Random Walk scheme, i.e. picking n gateway nodes connected to one randomly chosen overlay at each step. This strategy, henceforth referred to as *n-Random Walk*, helps limit the number of messages in the system, and is the one that we have tested in our performance evaluation of Synapse in Section 3. Other examples of routing strategies may include *n-Flooding* or *Direct Routing* to specific overlay networks, where a SYNAPSE_REQUEST message embeds the list of targeted netIDs, and the message is routed along the gateway nodes until the desired overlays are reached.

Gateway Node Discovery. In order to reach beyond its own overlays, a synapse node which joins the network needs to discover gateway nodes connected to overlays other than its own. The most effective way of discovering gateway nodes would be *Message Embedding*: each node embeds its list of its connected overlays in every message sent in the overlays, In this way, every other node routing the message can update its DOT with the embedded gateway information.

If, however, the presence of legacy nodes makes it impossible to alter the message packets in the overlay, an alternative mechanism would be *Active Notification*: gateway nodes routing a message in the overlay contact the message originator by sending a SYNAPSE_OFFER message with the list of their connected overlays. There could also be a third scenario in which, due to the overlay routing being, for example, fully recursive, the message originator is not known. In this case, a *Peer Exchange* mechanism serves the purpose of populating the DOT of each node.

3 Simulation and Experimental Results

In this section, we present the results obtained by running simulations within the OverSim-based Synapse simulator, as well as those obtained from the deployment of Synapse on the Grid'5000 platform.

OverSim Implementation of Synapse. In order to be able to handle multiple overlay networks of different types, we have modified the OverSim simulator to support dynamic run-time instantiation of overlay modules and the interaction of existing overlay modules in OverSim with the Synapse-controller module. A more detailed description of these changes can be found in [17].

Simulation Settings. All of the simulations were run on 2000 nodes, clustered into an equal number of Chord and Kademlia sub-overlays. All of the nodes were treated as either synapse or gateway nodes, and no legacy nodes were present. The main purpose of our simulations was to test the reachability of each of the overlays, while varying the granularity of the network (i.e. the number of different overlays (o), given the same overall amount of nodes (n)), the number of gateway nodes present in each of the overlays (g_o), and the connection degree of gateway nodes (i.e. the number of different overlays a gateway node is connected to (d)). As the idea was to estimate a lower bound of system performance, we have adopted the simplest and least demanding routing strategy for synapse requests: a stateless 1-Random-Walk, as described in Section 2. Finally, the TTL has been set to 8, for all of the simulations.

The tests consisted of inserting random keys throughout the entire system, and performing lookups for said keys, by a different node, not necessarily a member of the same sub-overlay in which the key is present. All replication within the sub-overlays has been disabled in order to create the most challenging conditions, and produce metrics as correlated as possible. We have tested different scenarios, without churn, to evaluate the topology built by the node discovery process, and with high churn, i.e. with a very short node lifetime, to test it in extreme conditions, such as those of a mobile application. In all of the simulations, the connection degree was equal for all gateway nodes. However, the percentage of gateway nodes and their interconnection degree have been correlated to guarantee the minimum number of gateway nodes-per-overlay to have a connected topology across all sub-overlays, without leaving any sub-overlay isolated due to, possibly, a lack of gateway nodes connected to it.

Topology Construction. Topologies have been created statically, using n, o, d, and, depending on the simulation scenario, either the percentage of gateway nodes-per-overlay, or the overall percentage of gateway nodes in the system. Two algorithms were used to generate topologies:

1. FIT – the topology is constructed to be fully-interconnected, in the sense that from any overlay there exists a path through gateway nodes of the system to any other overlay. This requires at least $\lceil \frac{o-1}{d-1} \rceil$ gateway nodes to be present in the system, and is accomplished using an algorithm described at [2];
2. RAT – The topology is constructed with fully random assignments of overlays to gateway nodes, using a uniform distribution over the o overlays.

Simulation 1: Effects of System Granularity. Figure 2 shows the effect that system granularity (i.e. the number of sub-overlays) has on the general system exhaustiveness. We have simulated both a churn-less environment and one with high churn, to test the topology itself, as well as its resilience to extreme conditions. Figure 2a and Figure 2b compare a completely random topology vs. the one in which exhaustive connectivity has been forced. It is remarkable that the performances are substantially equivalent, suggesting that, in fact, a gateway topology can generally be built with just a partial knowledge of the system, by a simple random selection of overlays. Even with 200 overlays, the routing has proven to be exhaustive, reaching every sub-overlay, and suggesting that building a clustered overlay network is a feasible solution. Lower

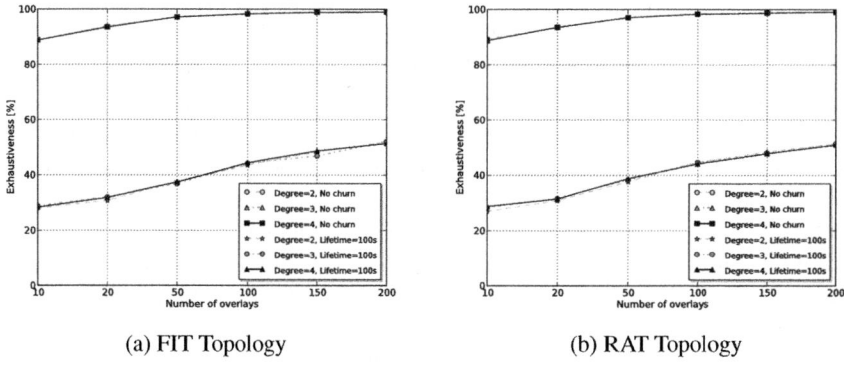

(a) FIT Topology   (b) RAT Topology

Fig. 2. Effects of system granularity, with and without churn

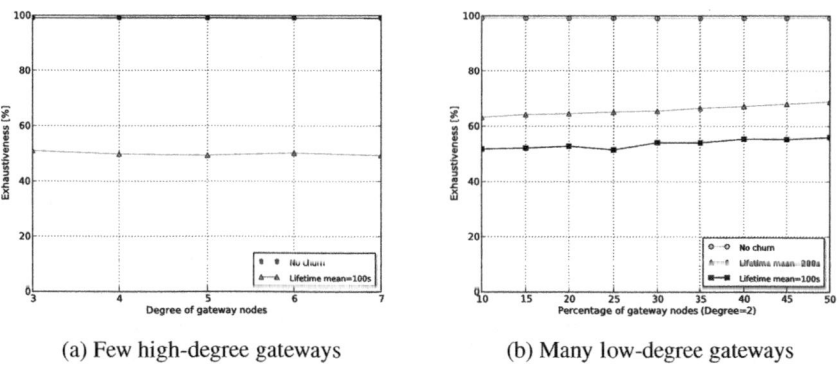

(a) Few high-degree gateways   (b) Many low-degree gateways

Fig. 3. Performance comparison for different gateway topologies

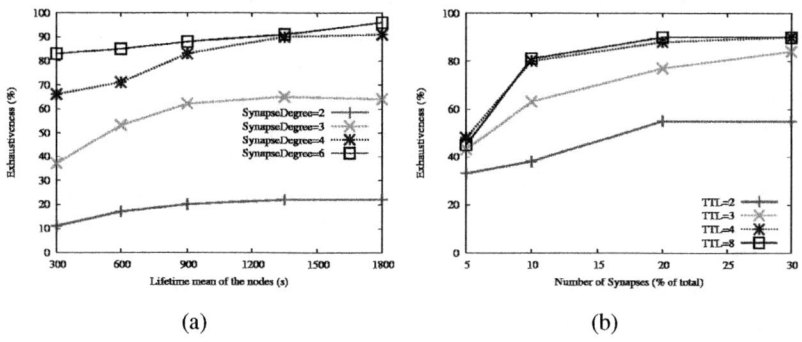

(a)   (b)

Fig. 4. Results from experiments run on Grid'5000

exhaustiveness with lower granularity is explained by the fact that, with having a higher number of edges for each overlay, there is a higher probability of loops being present, leading, with this simplest routing strategy, to requests bouncing back to their original overlay, an effect which can easily be avoided with a stateful routing strategy.

Simulation 2: Configuration of Gateway Nodes. Since maintaining a connection to multiple overlays is a costly operation, in this experiment we have tested the effectiveness of two opposite scenarios, one with very few gateway nodes maintaining a high degree of connectivity (much like a super-peer structure), and a second one, in which an increasing number of gateway nodes maintains a connectivity as low as possible (degree 2). It is worth noting that, despite the high connectivity degree, the gateway nodes in the first scenario were not exempted from churning. Figure 3 shows the system performance in the two scenarios. Interestingly enough, a decrease from degree 6 to degree 3 (Figure 3a) does not bring any visible decrease in performance, neither with nor without churn, partly due to the simple routing strategy adopted, and it is an aspect that can be taken into account when designing a system by explicitly deploying synapse-gateways. In the second scenario (Figure 3b), on the other hand, the increase of gateway nodes brings a slight increase in the exhaustiveness under churn, which suggests a possible strategy to handle situations of sudden churn in a system, by having most of the nodes immediately increase their connectivity degree by 1.

Experiments on Grid'5000. In order to evaluate the behavior of Synapse within a real-world environment, we have developed a Java implementation of the Synapse protocol, which we have used to perform experiments on the Grid'5000 platform, which aims at providing an experimentation testbed to study large-scale parallel or distributed systems, and comprises thousands of interconnected computers across numerous sites in France. In all of the experiments performed, we have used 1000 nodes, distributed over 10 Chord and 10 Kademlia overlays, interconnected via the Synapse protocol.

In the first experiment, we have investigated the exhaustiveness of the interconnected systems under different mean lifetimes of the nodes and different degrees of connectivity of synapse nodes. We have placed an emphasis on high-churn-rate conditions (when the mean lifetime of the nodes is low), which should be observable in the near future, in overlay networks in which peers need not only be desktops and laptops, but also Internet TV and mobile devices, which are expected to join and leave the network at high frequency. In order to generate this high churn rate of nodes in the systems, we have used the Pareto distribution. The overall percentage of synapse nodes was fixed to 20% of the overall number of nodes, while the TTL value was fixed to 8, in all of the cases. The results obtained from this experiment are shown in Figure 4a, from which we can notice that, for a fixed degree of connectivity, the Synapse protocol is fairly resilient for values of the mean lifetime above 900s, and less resilient for lower values. However, in order to achieve a sufficient level of exhaustiveness, it is necessary to increase the degree of connectivity of synapse nodes to at least 4, for mean lifetime values above 900s, or to at least 6, for mean lifetime values below 600s.

In the second experiment, we have once again investigated the exhaustiveness of the interconnected systems, this time while varying the percentage of synapse nodes and the TTL. The degree of connectivity of synapse nodes has been fixed to 4, and the churn rate of the nodes to 1800s. The results obtained from this experiment are shown in Figure 4b, from which it can be noticed that the exhaustiveness significantly increases when the TTL is increased from 2 to 4, but remains the same as the TTL is increased from 4 to 8, giving rise to the conclusion that a TTL of 4 is efficient enough when interconnected networks of this scale are concerned (20 networks, 1000 nodes overall) Another inference which can be made from Figure 4b is that having 20% of overall nodes as synapses will result in sufficient exhaustiveness for this scale of interconnected overlays, as there is an obvious rise in exhaustiveness accompanying the increase of the number of synapse nodes from 5% to 10% and from 10% to 20%, while no significant rise occurs with a further increase of the number of synapses from 20% to 30%.

4 Conclusions and Further Work

In this paper, we have given a brief overview and presented the first evaluation of the Synapse Framework, the purpose of which is to enable the design of distributed applications based on multiple interconnected overlays, as well as to facilitate easier interconnection of already deployed overlays. The protocol has been developed in the OverSim overlay simulator, which has been modified to support multiple overlay types at run-time, and a Java client has been deployed and tested on the Grid'5000 platform. Simulations and experimental results suggest that the framework is suitable for providing efficient support for federated topologies with limited costs, in terms of messages and node connectivity. As we have just begun scratching the surface of all the possibilities offered by such an approach, our future work includes additional mathematical modeling of the protocol, an adaptation to unstructured overlays as well and further extensive testing of all the routing strategies under different system parameters, in order to be able to accurately quantify messaging overhead, resilience to churn and data consistency. Furthermore, we will aim to define a mechanism which would guarantee only a minimum level of interconnection between different overlays, i.e. that there is a constant presence of only a minimal number of gateway nodes within the overlays.

Acknowledgments. The authors warmly thank Erol Gelenbe for many precious discussions, and Riccardo Loti for careful reading.

References

1. Bittorrent website, http://www.bittorrent.com
2. Synapse website, http://www-sop.inria.fr/teams/lognet/synapse
3. Baumgart, I., Heep, B., Krause, S.: OverSim: A scalable and flexible overlay framework for simulation and real network applications. In: Proceedings of the IEEE P2P 2009, pp. 87–88 (2009)
4. Castro, M., Costa, M., Rowstron, A.: Peer-to-peer overlays: Structured, unstructured, or both? Technical Report MSR-TR-2004-73, Microsoft Research, Cambridge, UK (2004)
5. Cheng, L.: Bridging distributed hash tables in wireless ad-hoc networks. In: Proc. of GLOBECOM 2007 (2007)

6. Cheng, L., Ocampo, R., Jean, K., Galis, A., Simon, C., Szabó, R., Kersch, P., Giaffreda, R.: Towards Distributed Hash Tables (De)Composition in Ambient Networks. In: State, R., van der Meer, S., O'Sullivan, D., Pfeifer, T. (eds.) DSOM 2006. LNCS, vol. 4269, pp. 258–268. Springer, Heidelberg (2006)

7. Datta, A., Aberer, K.: The challenges of merging two similar structured overlays: A tale of two networks. In: Proc. of EuroNGI 2006 (2006)

8. Ganesan, P., Krishna Gummadi, P., Garcia-Molina, H.: Canon in g major: Designing dhts with hierarchical structure. In: Proc. of ICDCS 2004 (2004)

9. Garcés-Erice, L., Biersack, E.W., Felber, P., Ross, K.W., Urvoy-Keller, G.: Hierarchical Peer-to-Peer Systems. In: Kosch, H., Böszörményi, L., Hellwagner, H. (eds.) Euro-Par 2003. LNCS, vol. 2790, pp. 1230–1239. Springer, Heidelberg (2003)

10. Jimenez, R., Osmani, F., Knutsson, B.: Connectivity properties of mainline bittorrent dht nodes. In: Proc. of IEEE P2P 2009 (2009)

11. Kwon, M., Fahmy, S.: Synergy: an overlay internetworking architecture. In: Proc. of ICCCN 2005 (2005)

12. Liquori, L., Tedeschi, C., Vanni, L., Bongiovanni, F., Ciancaglini, V., Marinković, B.: Synapse: A Scalable Protocol for Interconnecting Heterogeneous Overlay Networks. In: Crovella, M., Feeney, L.M., Rubenstein, D., Raghavan, S.V. (eds.) NETWORKING 2010. LNCS, vol. 6091, pp. 67–82. Springer, Heidelberg (2010)

13. Thau Loo, B., Huebsch, R., Stoica, I., Hellerstein, J.M.: The Case for a Hybrid P2P Search Infrastructure. In: Voelker, G.M., Shenker, S. (eds.) IPTPS 2004. LNCS, vol. 3279, pp. 141–150. Springer, Heidelberg (2005)

14. Shafaat, T.M., Ghodsi, A., Haridi, S.: Dealing with network partitions in structured overlay networks. Peer-to-Peer Networking and Applications 2(4) (2009)

15. Tan, G., Jarvis, S.A.: Inter-overlay cooperation in high-bandwidth overlay multicast. In: Proc. of ICPP 2006 (2006)

16. Xu, Z., Min, R., Hu, Y.: Hieras: A dht based hierarchical p2p routing algorithm. In: Proc. of ICPP 2003 (2003)

17. Ciancaglini, V., Liquori, L., Ngo Hoang, G.: Towards a Common Architecture to Interconnect Heterogeneous Overlay Networks. In: Proc of ICPADS 2011 (2011)

Dynamic Pricing Coalitional Game for Cognitive Radio Networks*

Yong Xiao and Luiz A. DaSilva

CTVR, Trinity College Dublin, Ireland
{yongx,dasilval}@tcd.ie

Abstract. We consider a hierarchical game theoretic model for cognitive radio (CR) networks in which primary users (PU) set the price to charge secondary users (SU) for accessing the licensed spectrum and SUs optimize their transmit powers according to the price imposed by PUs. Pricing strategies can be tailored to steer SUs to a Stackelberg equilibrium. We establish a coalition formation game framework to study the possible cooperation among PUs. In our framework, the PUs who can detect the same SUs form a coalition to select the pricing function as long as each member of the coalition is allocated a fair share of the payoff. We show that allowing all PUs to cooperatively decide the price for every SU is generally not the optimal solution. We then propose a distributed algorithm that allows PUs to dynamically approach a unique and stable partition of the grand coalition, as well as a Stackelberg equilibrium point of the hierarchical game.

Keywords: Coalition formation, cognitive radio, Stackelberg game, game theory.

1 Introduction

Radio spectrum is generally regarded as a scarce resource. This motivates a new hierarchical network framework, taking advantage of cognitive radios (CR), in which the unlicensed users, called secondary users (SU), can learn from their surrounding environment and intelligently decide how to opportunistically utilize the spectrum licensed to the spectrum owners, called primary users (PU). Different CR network models have been proposed based on the tolerance of the PUs to interference caused by the SUs. More specifically, by assuming PUs cannot tolerate any interference caused by SU networks, temporal spectrum sharing (TSS) [1] was proposed to allow each SU to detect temporal vacancy of PUs. By assuming that each PU can only tolerate a small increase in the interference caused by SUs, spatial spectrum sharing (SSS) [2–4] was studied. In this system, SUs can send signals over the licensed spectrum as long as the resulting

---

* This material is based upon works supported by the Science Foundation Ireland under Grant No. 10/IN.1/I3007.

Z. Becvar et al. (Eds.): NETWORKING 2012 Workshops, LNCS 7291, pp. 19–26, 2012.

interference power is lower than the maximum tolerable level, called the interference temperature limit. In SSS-based CR networks, how to control the transmit powers of SUs to satisfy the power constraints imposed by PUs is an important problem.

In this paper, we establish a Stackelberg game-based hierarchical framework in which PUs have priority in using the licensed spectrum, as well as setting the prices for SUs, and each SU tries to improve its performance according to the prices imposed by PUs. A possible scenario for our setting could be a heterogeneous cellular network in which independent mobile devices (SUs) dynamically select among multiple service providers (PUs) and pay a corresponding price. To study the possible pricing competition and cooperation among PUs, we propose a novel coalition formation game framework for the PUs, referred to as the dynamic pricing coalitional game. In this framework, a collection of coalitions is formed by different groups of PUs to decide the prices their nearby SUs will be charged. To study the effects of the PU cooperations on the performance of SUs, we fit the proposed dynamic pricing coalitional game into a hierarchical framework. We prove that allowing all PUs to cooperatively decide the price for every SU is generally not the optimal choice. This is different from previous work in coalitional game-based wireless networks which typically simplifies the system model by neglecting the cooperation costs [5]. We then focus on developing effective methods for PUs to search for the optimal partitions of the grand coalition. It is observed that the coalitions formed among PUs to decide the prices of different SUs are always correlated and it is generally not feasible or efficient to use the exhaustive coalition searching method [6]. In addition, as observed in [7], finding a low-complexity algorithm for coalition formation games with overlapping coalitions is generally difficult because of the combinatorial complexity order caused by distributing the benefits of each member among multiple coalitions. In this paper, we propose a simple distributed coalition formation algorithm which allows PUs to form a unique, stable partition without any knowledge of channel conditions experienced by SUs.

2   System Model and Basic Game Setup

Let the sets of J PUs and K SUs be $\mathcal{J} = \{P_1, P_2, ..., P_J\}$ and $\mathcal{K} = \{S_1, S_2, ..., S_K\}$, respectively. We assume that SUs use OFDMA and each SU is pre-allocated a frequency band for its transmissions. We assume the PUs and SUs belong to different networks and have no a priori knowledge of each other's channel state information or transmit powers. In a practical system, the PUs can only interact with the SUs they can detect. In this paper, we assume that each PU will first use an energy detector to determine the presence of SUs in each frequency band. As observed in [8], PUs can only detect an SU if the received SNR from this SU is larger than a threshold, called the *SNR wall*. Let us define the SNR wall for P_j as $\underline{q}_{P_j}$, i.e., P_j can detect the existence of a SU S_k if $h_{jk}w_{S_k} > \underline{q}_{P_j}$, where h_{jk} is the channel gain between S_k and P_j and w_{S_k} is the transmit power of S_k.

Another constraint for the SUs is the interference temperature limit of PUs. We assume each PU P_j imposes an interference temperature limit $\bar{q}_{P_j}$ for $\bar{q}_{P_j} > \underline{q}_{P_j}$ $\forall P_j \in \mathcal{J}$ in each frequency band of SUs, i.e., $h_{jk} w_{S_k} < \bar{q}_{P_j}$. Hence the power constraint for an SU S_k is defined as $h_{jk} w_{S_k} \leq \min_{P_j \in \mathcal{J}} \left\{ \bar{q}_{P_j} \right\}$.

Let us introduce a hierarchical game theoretic framework in which players of the game are the PUs (leader) who have priority in using the spectrum, and the SUs (follower) who can access the licensed spectrum by paying a certain "price". Prices are used by the PUs to distributedly regulate the transmit powers of the SUs, so as to achieve an optimal trade-off between spectrum utilization and the interference to the PU network. Let the subset of PUs who can detect the existence of S_k be $\mathcal{C}_{P[S_k]}$ and the subset of SUs who are visible to P_j be $\mathcal{C}_{P_j[S]}$. We define the payoff of S_k to be

$$\varpi_{S_k} \left(w_{S_k}, \boldsymbol{\beta}_{P[S_k]} \right) = \alpha_{S_k} \log \left(1 + g_{S_k} w_{S_k} \right) - \boldsymbol{\beta}_{P[S_k]} \boldsymbol{h}_{\bullet k} w_{S_k}, \tag{1}$$

where g_{S_k} is the ratio of the channel gain between the kth secondary sender-to-receiver pair to the additive interference power received by S_k, α_{S_k} is a positive constant, $\boldsymbol{\beta}_{P[S_k]} = \left(\beta_{P_j[S_k]} \right)_{P_j \in \mathcal{C}_{P[S_k]}}$, $\beta_{P_j[S_k]}$ is the pricing coefficient of P_j charged to S_k and $\boldsymbol{h}_{\bullet k} = [h_{1k}, h_{2k}, ..., h_{Jk}]^{\dagger}$, $\dagger$ denotes the transpose of a matrix. In this paper, we assume each SU can use optimal power control by solving the following problem,

$$w^*_{S_k} \left(\boldsymbol{\beta}_{P[S_k]} \right) = \arg \max_{w_{S_k}} \varpi_{S_k} \left(w_{S_k}, \boldsymbol{\beta}_{P[S_k]} \right). \tag{2}$$

Solving (2), we can obtain the optimal transmit power of each SU below,

$$w^*_{S_k} \left(\boldsymbol{\beta}_{P[S_k]} \right) = \alpha_{S_k} \left(1/u_{S_k}(\boldsymbol{\beta}_{P[S_k]}) - 1/\delta_{S_k} \right)^+, \tag{3}$$

where $(x)^+ = \max\{x, 0\}$, $u_{S_k} \left(\boldsymbol{\beta}_{P[S_k]} \right) = \boldsymbol{\beta}_{P[S_k]} \boldsymbol{h}_{\bullet k}$ and $\delta_{S_k} = \alpha_{S_k} g_{S_k}$. It is observed that the transmit power of an SU can only be non-zero if $u_{S_k} \left(\boldsymbol{\beta}_{P[S_k]} \right) < \delta_{S_k}$. Let us define the payoff of P_j as follows,

$$\varpi_{P_j}(\boldsymbol{w}_S, \boldsymbol{\beta}_{P_j[S]} | \mathcal{C}_{P_j[S]}) = \sum_{S_k \in \mathcal{C}_{P_j[S]}} \left(\pi_{P_j[S_k]} - \theta_{P_j[S_k]} \left(\mathcal{C}_{P[S_k]} \right) \right), \tag{4}$$

where $\boldsymbol{\beta}_{P_j[S]} = \left(\beta_{P_j[S_k]} \right)_{S_k \in \mathcal{C}_{P_j[S]}}$ and $\pi_{P_j[S_k]} = \beta_{P_j[S_k]} h_{jk} w_{S_k}$ is the revenue obtained by P_j from S_k. We assume for each PU the revenue obtained from different SUs is independent. $\theta_{P_j[S_k]} \left(\mathcal{C}_{P[S_k]} \right) \geq 0$ is the cooperation cost of P_j when it joins a coalition $\mathcal{C}_{P[S_k]}$. $\theta_{P_j[S_k]} \left(\mathcal{C}_{P[S_k]} \right) = 0$ if P_j does not belong to a coalition to decide the price of S_k, i.e., $\mathcal{C}_{P[S_k]} = \emptyset$, or is the only element in a coalition to provide spectrum for S_k, i.e., $\mathcal{C}_{P[S_k]} = \{P_j\}$. If P_j is involved in a multiple-PU coalition, $\theta_{P_j[S_k]} \left(\mathcal{C}_{P[S_k]} \right)$ should be a positive value related to the transmit power and/or the time spent in sending and receiving cooperation-related information between the member PUs in the coalition $\mathcal{C}_{P[S_k]}$ [6].

In this paper, we consider the pricing coalitional game in which the PUs within one coalition only care about their payoff sum, which will be divided among all the members according to an appropriate fairness criterion. We hence can regard all PUs in $\mathcal{J}$ as one, labeled as $P_{\mathcal{J}}$, with payoff $\varpi_{\mathcal{J}}\left(w_S^*, \beta_{P_j[S]}\right) =$ $\sum_{P_j \in \mathcal{J}} \varpi_{P_j}\left(w_S^*, \beta_{P_j[S]}\right)$ for $\beta_{\mathcal{J}} = \left(\beta_{P_j[S_k]}\right)_{S_k \in \mathcal{K}, P_j \in \mathcal{J}}$. One of the main objectives for our hierarchical game framework is to find an equilibrium point, called the Stackelberg Equilibrium (SE), for our combined game.

3   Game Theoretic Analysis and Coalition Formation Algorithm

Let us formally define the concept of the coalition below.

Definition 1. *[9, Chapter 9] A coalition $\mathcal{C}$ is a non-empty sub-set of the total set of players $\mathcal{J}$, i.e., $\mathcal{C} \subseteq \mathcal{J}$. We refer to the coalition of all the players as the grand coalition $\mathcal{J}$. A coalitional game is defined by the pair $(\mathcal{J}, v)$ where v is called the characteristic function, which assigns a number $v(\mathcal{C})$ to every coalition $\mathcal{C}$ and $v(\emptyset) = 0$. Here $v(\mathcal{C})$ quantifies the worth of a coalition $\mathcal{C}$. A coalitional game is said to be super-additive if for any two disjoint coalitions $\mathcal{C}^1$ and $\mathcal{C}^2$, $\mathcal{C}^1, \mathcal{C}^2 \subset \mathcal{J}$, we have $v(\mathcal{C}^1 \cup \mathcal{C}^2) \geq v(\mathcal{C}^1) + v(\mathcal{C}^2)$.*

We have the following remark about the stability of the grand coalition for our game.

Remark 1. The grand coalition is often unstable for a multi-user CR network.

Let us illustrate this through an example. Suppose that all SUs use the same transmit powers and three PUs with equal SNR wall and interference temperature limit are located in a linear network as shown in Figure 1. Each PU can only detect its nearby SUs (we represent the detection area of each PU as a shadowed circle in Figure 1). Thus, it can be observed that P_3 can only obtain positive payoff from the three closest SUs S_2, S_3 and S_5 and cannot obtain any revenue from the farthest SUs S_1 and S_4. Because the distance between PUs P_3 and P_1 is large, the cooperation cost for forming a coalition is large too. In other words, if the cooperation costs of P_3 to charge S_2 and S_3 is larger than the payoff obtained from S_2, S_3 and S_5, P_3 will have no incentive to join the grand coalition but will only form a coalition with P_2 to charge S_2, S_3 and S_5. The above remark can be easily extended to a general CR network with SUs and PUs randomly located in a large area.

Let us now consider the possible pricing coalition formation among PUs. As a motivation example, we consider Figure 1 again. It is observed that the different coalitions of PUs to decide the prices charged to different SUs may not be independent. For example, in Figure 1, P_2 should cooperate with P_1

on deciding the price changed to S_1 and also cooperate with P_3 on choosing the price charged to S_3. Another observation is that the cooperation between two disjoint coalitions may not always improve the payoff sum. Assume that the channel gains between S_2 and three PUs (P_2, P_1 and P_3) satisfy

$$\underline{q}_{P_2}/h_{22} < \underline{q}_{P_1}/h_{12} < \underline{q}_{P_3}/h_{32}, \quad \max_{j\in\{1,2,3\}}\left\{\underline{q}_{P_j}/h_{j2}\right\} < \min_{j\in\{1,2,3\}}\left\{\overline{q}_{P_j}/h_{j2}\right\} \text{ and}$$

$$v\left(\mathcal{C}^1_{P[S_2]}\cup\mathcal{C}^2_{P[S_2]}\right) - v\left(\mathcal{C}^1_{P[S_2]}\right) - v\left(\mathcal{C}^2_{P[S_2]}\right) > \sum_{P_j\in\mathcal{C}^1_{P[S_2]}\cup\mathcal{C}^2_{P[S_2]}} \theta_{P_j[S_k]}(\mathcal{C}^1_{P[S_2]}\cup$$

$$\mathcal{C}^2_{P[S_2]}) - \sum_{P_j\in\mathcal{C}^1_{P[S_2]}} \theta_{P_j[S_k]}(\mathcal{C}^1_{P[S_2]}) - \sum_{P_j\in\mathcal{C}^2_{P[S_2]}} \theta_{P_j[S_k]}(\mathcal{C}^2_{P[S_2]}) \text{ where } \mathcal{C}^1_{P[S_2]} \text{ and } \mathcal{C}^2_{P[S_2]}$$

are two disjoint sub-sets of $\{P_1, P_2, P_3\}$. In this case, If P_1 and P_2 form a coalition to charge S_2, the resulting payoff sum is always larger than that without cooperation. However, this result does not hold when P_2 and P_3 cooperate without P_1, i.e., the payoff sum is $\sum_{i\in\{2,3\}} \varpi_{P_i[S_2]} = \beta_{P_2[S_2]}h_{22}\underline{q}_{P_1}/h_{12} - \sum_{i=\{2,3\}} \theta_{P_i}(\{P_2, P_3\})$ which is always worse than the payoff sum without cooperation. To solve the

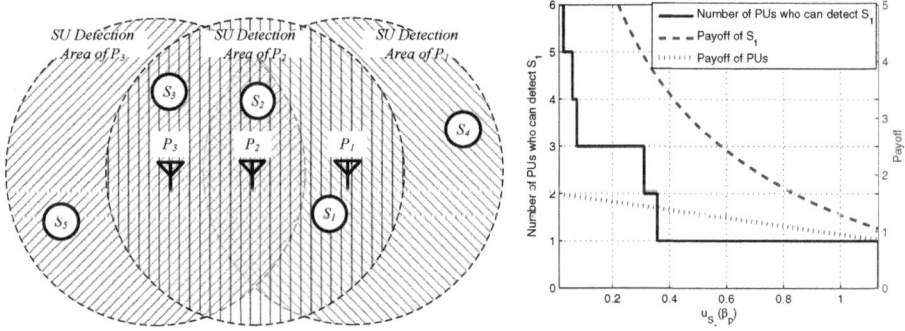

Fig. 1. SU detection area of three PUs   **Fig. 2.** Coalition formation of PUs

first issue, we seek to convert all the correlated coalitions into independent ones as follows. It is observed that the payoff function of each PU P_j in (4) is the summation of all the payoff functions related to its detectable SUs. Since these payoff functions are independent, maximizing the payoff of each PU P_j corresponds to maximizing the payoff of P_j earned from each detectable SU. We hence can separate the payoff of the PU P_j into different independent parts according to different SUs. In this way, during the rest of this paper, we only need to focus on a pricing coalitional game in one frequency band in which a set of PUs, denoted as $\mathcal{C}_{P[S_k]} = \left\{P_j : h_{jk}w_{S_k} \geq \underline{q}_{P_j}\right\}$, cooperates with one another in deciding the price charged to a SU S_k. To solve the second problem, we rearrange the labeling sequence of the PUs in $\mathcal{C}_{P[S_k]}$ by $\{P_{\tilde{1}}, P_{\tilde{2}}, ..., P_{\widetilde{|\mathcal{C}_{P[S_k]}|}}\}$ where $\underline{q}_{P_{\widetilde{j-1}}}/h_{\widetilde{j-1}k} < \underline{q}_{P_{\tilde{j}}}/h_{\tilde{j}k} < \underline{q}_{P_{\widetilde{j+1}}}/h_{\widetilde{j+1}k}$ for $0 < j < |\mathcal{C}_{P[S_k]}| - 1$. We say the PUs are *sequential* if their rearranged labels are consecutive, i.e., $P_{\widetilde{j-1}}, P_{\widetilde{j}}, ..., P_{\widetilde{j+l}}$ are sequential. We say one set is *sequential* if all the elements in this set are sequential. We say two or more disjoint sets are *sequential* if each of these sets are sequential and the union of these sets is sequential too, i.e., $\mathcal{C}^1 = \{P_{\widetilde{1}}, ..., P_{\widetilde{j}}\}$ and $\mathcal{C}^2 = \{P_{\widetilde{l+1}}, ..., P_{\widetilde{j}}\}$ for $1 < l < j$ are sequential. We have the following property for the proposed game.

Proposition 1. *Assume the interference temperature limit is always satisfied. Suppose two disjoint coalitions $\mathcal{C}^1_{P[S_k]}$ and $\mathcal{C}^2_{P[S_k]}$ for $\mathcal{C}^1_{P[S_k]}, \mathcal{C}^2_{P[S_k]} \subset \mathcal{C}_{P[S_k]}$ satisfy the following conditions,*

1) $P_{\widetilde{1}} \in \mathcal{C}^1_{P[S_k]} \cup \mathcal{C}^2_{P[S_k]}$,
2) $\mathcal{C}^1_{P[S_k]} \cup \mathcal{C}^2_{P[S_k]}$ *is sequential.*
3) $v\left(\mathcal{C}^1_{P[S_k]} \cup \mathcal{C}^2_{P[S_k]}\right) - \sum_{n=1,2} v\left(\mathcal{C}^n_{P[S_k]}\right) >$

$$\sum_{P_j \in \mathcal{C}^1_{P[S_k]} \cup \mathcal{C}^2_{P[S_k]}} \theta_{P_j[S_k]}\left(\mathcal{C}^1_{P[S_k]} \cup \mathcal{C}^2_{P[S_k]}\right) - \sum_{n=1,2} \sum_{P_j \in \mathcal{C}^n_{P[S_k]}} \theta_{P_j[S_k]}\left(\mathcal{C}^n_{P[S_k]}\right) > 0$$

Then, $\mathcal{C}^1_{P[S_k]}$ and $\mathcal{C}^2_{P[S_k]}$ are super-additive.

Proof. The above result can be obtained from the definitions of super-additive and sequential coalition. We hence omit the details for space limitation. □

In this paper, we refer to the constrained coalitional game with all the subsets of member PUs of a coalition $\mathcal{C}_{P[S_k]}$ satisfying the above conditions as a *sequential coalitional game*. Let us define the preference sign in comparing different partitions as follows.

Definition 2. *Let $\mathcal{S} = \{\mathcal{S}^1, \mathcal{S}^2, ..., \mathcal{S}^l\}$ and $\mathcal{T} = \{\mathcal{T}^1, \mathcal{T}^2, ..., \mathcal{T}^m\}$ be two partitions of $\mathcal{J}$ with $\bigcup_{i \in \{1,2,...,l\}} \mathcal{S}^i = \bigcup_{j \in \{1,2,...,m\}} \mathcal{T}^j = \mathcal{J}$. Defining a comparison relation $\rhd$, $\mathcal{S} \rhd \mathcal{T}$ means that the way $\mathcal{S}$ partitions $\mathcal{J}$ is preferable to the way $\mathcal{T}$ partitions $\mathcal{J}$. In this paper, we consider Pareto order, i.e., $\mathcal{S} \rhd \mathcal{T}$ means $\varpi^{\mathcal{S}}_{P_j} \geq \varpi^{\mathcal{T}}_{P_j}, \forall P_j \in \mathcal{S}, \mathcal{T}$ with at least one strict inequality ($>$) for a PU P_j.*

We say a partition $\mathcal{S} = \{\mathcal{S}^1, \mathcal{S}^2, ..., \mathcal{S}^l\}$ of $\mathcal{J}$ is *stable* if no group of players has an incentive to leave $\mathcal{S}$. We assume PUs can use common knowledge or previous observation about SU networks to estimate the approximate upper bound of pricing coefficient $\bar{\beta}$, i.e., if $\beta_{P_j[S_k]} = \bar{\beta} \, \forall P_j \in \mathcal{J}, S_k \in \mathcal{K}$, no SUs can afford the price of PUs. We present the detailed description of the algorithm as follows.

---

1) *Initialization:* Set $\mathcal{C}_{P[S_k]}(0) = \emptyset$ and $\mathcal{C}_{P_j[S]}(0) = \emptyset$. Each PU P_j broadcasts a high pricing coefficient vector $\boldsymbol{\beta}_{P_j[S]}(0) = [\beta_{P_j[S_1]}(0), \beta_{P_j[S_2]}(0), ..., \beta_{P_j[S_K]}(0)]$ where $\beta_{P_j[S_k]}(0) \geq \bar{\beta}^1$.

---

[1] Note that, at the beginning of each iteration, PUs need to pre-set the prices for each frequency band of SUs without knowing how many SUs can afford the price. Hence, we abuse the notation and use $\beta_{P_j[S_k]}(t)$ to denote the price that P_j sets for use of the frequency band of S_k even if $w^*_{S_k} = 0$.

2) *Coalition Formation*: Receiving $\beta_{P[S_k]}(t)$, each SU S_k sets its transmit power $w^*_{S_k}$. At iteration t, if a PU P_j cannot detect any SUs, i.e., $\mathcal{C}_{P_j[S]} = \emptyset$, then jump to Step 3) directly. If a PU P_j detects the transmission of at least one SU, it sends the list $\mathcal{C}_{P_j[S]}(t)$ to other PUs for possible cooperation. If $\mathcal{C}_{P[S_k]} \neq \emptyset$, all PU $P_j \in \mathcal{C}_{P[S_k]}(t)$ will jointly decide $u_{S_k}\left(\beta_{P[S_k]}(t)\right)$ for S_k.

3) *Dynamic Coalition Updating*: At the end of iteration t, if $\mathcal{C}_{P_j[S]}(t) = \emptyset$, P_j will update the price $\beta_{P_j[S]}(t) = \beta_{P_j[S]}(t-1) - \epsilon$ for all frequency bands. If $\mathcal{C}_{P_j[S]}(t) \neq \emptyset$, P_j will jointly collaborate with other PUs $P_i \in \mathcal{C}_{P[S_k]} \, \forall \, S_k \in \mathcal{C}_{P_j[S]}$ to update the price $u_{S_k}\left(\beta_{P[S_k]}(t+1)|\mathcal{C}_{P[S_k]}(t+1)\right) = u_{S_k}\left(\beta_{P[S_k]}(t) - \epsilon|\mathcal{C}_{P[S_k]}(t)\right)$ for $S_k \in \mathcal{C}_{P_j[S]}$. In addition, P_j will also update the price $\beta_{P_j[S_k]}(t) = \beta_{P_j[S_k]}(t-1) - \epsilon, \forall \, S_k \notin \mathcal{C}_{P_j[S]}(t), S_k \in \mathcal{K}$ for the frequency bands. Let $t = t+1$. Go to Step 2). If one PU $P_j \in \mathcal{C}_{P[S_k]}$ detects a higher than tolerable interference from S_k, the algorithm ends with solution $u_{S_k}\left(\beta_{P[S_k]}(t-1)\right)$, $\mathcal{C}_{P[S_k]}(t-1)$ and $\mathcal{C}_{P_j[S]}(t-1), \forall \, S_k \in \mathcal{K}, P_j \in \mathcal{J}$.

We have the following results about the above algorithm.

Theorem 1. *If the above algorithm terminates, either we have $\mathcal{C}_{P[S_k]} = \emptyset \, \forall S_k \in \mathcal{K}$, or we have*

1) *If $u_{S_k}\left(\beta^*_{P[S_k]}\right) = \dfrac{\alpha_{S_k}}{\min\limits_{P_i \in \mathcal{C}_{P[S_k]}} \{\bar{q}_{P_i}/h_{ik}\}+1/g_{S_k}} \, \forall S_k \in \mathcal{K}$ is satisfied, the resulting partition is unique, stable and $\triangleright$ maximal for a sequential coalitional game, and the resulting $\left(w^*_S, \beta^*_{\mathcal{J}}\right)$ is a pure strategy SE for the hierarchical game,*

2) *Else the resulting $\left(w^*_S, \beta^*_{\mathcal{J}}\right)$ is within an ϵ distance of an SE for the hierarchical game.*

Proof. Consider the possible coalition formed among PUs to decide the price charged to one SU S_k. In this case, the grand coalition $\mathcal{J}$ has been partitioned into two disjoint coalitions: $\mathcal{C}_{P[S_k]}$ and $\mathcal{C}^c_{P[S_k]} = \{P_j : h_{\tilde{j}k} w^*_{S_k} < \underline{q}_{P_j}\}$. Here we abuse the notation and use $\mathcal{C}$ to denote the partition of $\{\mathcal{C}, \mathcal{C}^c\}$ for $\mathcal{C} \cup \mathcal{C}^c = \mathcal{J}$. First, let us prove that the coalition formation in one iteration t of Step 2) in Algorithm 1 is unique, stable and $\triangleright$ maximal for a given pricing vector $\beta_{P[S_k]}$. From (3), it is observed that, if $\beta_{P[S_k]}(t)$ is fixed, the values of $u_{S_k}(\beta_{P[S_k]}(t))$, $w^*_{S_k}(\beta_{P[S_k]}(t))$ and the set of PUs who can detect S_k are fixed too. Thus $\mathcal{C}_{P[S_k]}$ is a unique result for the given $\beta_{P[S_k]}(t)$ which is also a unique vector for the chosen $\bar{\beta}$ and ϵ. Let us show that the resulting coalition formation is stable and $\triangleright$ maximal. Using the results of Proposition 1, we can prove that the resulting coalition $\mathcal{C}_{P[S_k]}(t)$ in iteration t has the following properties: P1) For any two disjoint sequential coalitions $\mathcal{C}^1 = \{P_{\tilde{1}}, ..., P_{\tilde{j}}\}$ and $\mathcal{C}^2 = \{P_{\widetilde{j+1}}, ..., P_{\tilde{l}}\}$ in $\mathcal{C}_{P[S_k]}$ such that $\tilde{j} = |\mathcal{C}^1|, \tilde{l} - \tilde{j} = |\mathcal{C}^2|$ and $\mathcal{C}^1 \cup \mathcal{C}^2 \subseteq \mathcal{C}_{P[S_k]}$, we have $\{\mathcal{C}^1 \cup \mathcal{C}^2\} \triangleright \{\mathcal{C}^1, \mathcal{C}^2\}$, P2) For any sequential coalition $\mathcal{C}^3 = \{P_{\tilde{1}}, ..., P_{\tilde{j}}\}$ such that $|\mathcal{C}^3| > |\mathcal{C}_{P[S_k]}|$ and $\mathcal{C}^3 \subseteq \mathcal{J}$, we have $\mathcal{C}_{P[S_k]} \triangleright \{\mathcal{C}^3\}$, P3) For any non-sequential coalition $\mathcal{C}^4$ such that $\mathcal{C}^4 \subseteq \mathcal{J}$, we have $\mathcal{C}_{P[S_k]} \triangleright \{\mathcal{C}^4\}$. By combining properties P1) - P3) and using the transitive, irreflexive and monotonic properties of $\rhd$ [10], we can claim that, for all partitions $\mathcal{C}^5 \neq \mathcal{C}_{P[S_k]}$ and $\mathcal{C}^5 \subseteq \mathcal{J}$, $\mathcal{C}_{P[S_k]} \rhd \mathcal{C}^5$ holds. From the above observation, we can claim that if a set of PUs $\Delta\mathcal{C}_{P[S_k]}(t)$ joins a coalition $\mathcal{C}_{P[S_k]}(t-1)$ in iteration t, following the Step 2) of Algorithm 1, it will have no incentive to leave the coalition $\mathcal{C}_{P[S_k]}(t)$. Let us consider the dynamic coalition updating step in Algorithm 1. The main effect of Step 3) in Algorithm 1 is to distributedly decrease the value of u_{S_k} until $w^*_{S_k}$ reaches its upper bound (the interference level increases to reach the interference temperature limit of at least u_{S_k} PU). Hence, the resulting $(w^*_{S_k}, \beta^*_{\mathcal{C}_{P[S_k]}})$ maximizes both the payoff of S_k and the payoff sum of $\mathcal{C}_{P[S_k]}$. This concludes our proof. □

In Figure 2, we show the size of a coalition $\mathcal{C}_{P[S_k]}$ and the payoff of S_k under different values of $u_{S_k}(\beta_{P[S_k]}|\mathcal{C}_{P[S_k]})$. It is observed that the size of the coalition as well as the payoffs of SUs decrease with $u_{S_k}(\beta_{P[S_k]}|\mathcal{C}_{P[S_k]})$. This verifies our observations that PUs can use $\beta_{P[S_k]}$ to control the partitions of the grand coalition, as well as the payoffs of SUs and PUs.

4 Conclusion

We build a hierarchical model for CR networks to investigate the emergence of pricing coalitions among PUs. We prove that the grand coalition of the coalitional game is generally not stable and hence we introduce a simple algorithm to allow PUs to distributedly form a unique and stable partition.

References

1. Zhao, Q., Sadler, B.M.: A survey of dynamic spectrum access. IEEE Signal Processing Magazine 24(3), 79–89 (2007)
2. Gastpar, M.: On capacity under receive and spatial spectrum-sharing constraints. IEEE Trans. Inform. Theory 53(2), 471–487 (2007)
3. Xiao, Y., Bi, G., Niyato, D.: Game theoretic analysis for spectrum sharing with multi-hop relaying. IEEE Trans. Wireless Commun. 10, 1527–1537 (2011)
4. Xiao, Y., Bi, G., Niyato, D.: A simple distributed power control algorithm for cognitive radio networks. IEEE Trans. Wireless Commun. 10, 3594–3600 (2011)
5. Mathur, S., Sankar, L., Mandayam, N.B.: Coalitions in cooperative wireless networks. IEEE J. Sel. Areas Commun. 26(7), 1104–1115 (2008)
6. Saad, W., Han, Z., Debbah, M., Hjorungnes, A.: A distributed coalition formation framework for fair user cooperation in wireless networks. IEEE Trans. Wireless Commun. 8(9), 4580–4593 (2009)
7. Zick, Y., Elkind, E.: Arbitrators in overlapping coalition formation games. In: AAMAS, pp. 55–62 (2011)
8. Tandra, R., Sahai, A.: SNR walls for signal detection. IEEE Journal of Selected Topics in Signal Processing 2(1), 4–17 (2008)
9. Myerson, R.B.: Game theory: analysis of conflict. Harvard University Press (1997)
10. Apt, K., Witzel, A.: A generic approach to coalition formation. International Game Theory Review 11(3), 347–367 (2009)

Evaluation of ON-OFF Schemes and Linear Prediction Methods for Increasing Energy Efficiency in Mobile Broadband Networks

Dario Sabella[1], Marco Caretti[1], William Tomaselli[1], Valerio Palestini[1],
Bruno Cendón[2], Javier Valino[2], Arturo Medela[2],
Yolanda Fernández[2], and Luis Sanchez[3]

[1] Telecom Italia, via Guglielmo Reiss Romoli, 274 - Turin, Italy
{dario.sabella,marco.caretti,guglielmo.tomaselli,
valerio.palestini}@telecomitalia.it
[2] TTI Norte, Avda. Albert Einstein 14, 39010, Santander, Spain
{bcendon,jvalino,amedela}@tst-sistemas.es,
yfernandez@ttinorte.es
[3] Universidad de Cantabria, Plaza de la Ciencia s/n, 39005, Santander, Spain
lsanchez@tlmat.unican.es

Abstract. Nowadays, energy efficiency has become a major issue in mobile networks operation. Due to the exponential rise in the number of wireless Internet-connected mobile devices reducing electrical energy consumption is not only a matter of showing environmental responsibility, but also of substantially reducing their operational expenditure. However, energy reduction cannot be pursued at any cost and appropriate service has to be supported. Among the diverse hardware and software solutions available, this paper focuses on the dynamic operation of cellular base stations, in which redundant base stations are switched off during periods of low traffic. Besides, we are also describing the use of prediction mechanisms in order to make a proper decision on when to take that action. The proposed schemes are assessed by means simulations using both theoretical and real load models.

Keywords: Green radio, energy saving, ON/OFF scheme, linear predictors, mobile communications, 3G, LTE, eNodeB, traffic load curves.

1 Introduction

The exponential growth in the number of devices connected to wireless networks makes the power consumption associated with their use also increase, which is against energy saving policies in place today. However, from the perspective of cellular network operators, reducing electrical energy consumption is not only a matter of being "green" and responsible, it is also very much an economically important issue: in fact, a significant portion of the operational expenditure (OPEX) of a cellular network goes to pay the electricity bill. From [1][2], it can be estimated that the mobile network OPEX for electricity globally is more than $10 billion dollars today. The key source

Z. Becvar et al. (Eds.): NETWORKING 2012 Workshops, LNCS 7291, pp. 27–34, 2012.
© IFIP International Federation for Information Processing 2012

of energy usage in cellular networks is the operation of Base Station (BS) equipment. It has been estimated that BSs contribute to 60–80 % of the total energy consumption [3]. Energy efficiency with respect to BSs has been considered in all stages of cellular networks, including hardware design and manufacture, deployment, and operation.

In this paper we will focus at the access network management level. So far, the different studies in the field of the switching on and off of BSs have focused on estimating the savings that could be achieved by dynamically adapting the operation of cellular network infrastructure to the required load. Initially load behaviour was simulated using linear, Gaussian or averaged models [3] [5]. In all of them the curves consists in monotonically decreasing or increasing functions and thus only a single low load "valley" appears. More recently, traffic profiles retrieved from a real network have been used for the same assessment [4], but still only the estimated savings are presented. The proposed scheme tries to dynamically minimize the number of active BSs to meet the traffic variation in the network, assuming that remaining active BSs are able to cope with the traffic that active users are generating. The optimum solution would require constantly mimicking the behaviour of the load profile and switching on and off cells as the traffic respectively increases or decreases. Since this way of operation demands complex self-organization issues, a simple scheme is designed, defining a priori the number of BSs that shall be put into idle mode, bearing in mind that the remaining sites could provide the required coverage to that specific area.

2   Reference Scenario

The main aim of the work presented in this paper is to create an intelligent network management mechanism to switch on and off BSs depending on the traffic load, reshaping at the same time the cell topology, always maintaining Quality of Service (QoS) and Quality of Experience (QoE), so the user will not notice any disruption in the service. Regarding cell geometry and positioning, the studies will consider typical grid with hexagonal geometry as it is the preferred and most conventional configuration. It is also important to clarify that tri-sectorial antennas will be used, in opposition to the omni-directional ones that were implemented in the previous studies, thus achieving results closer to reality. This option has been selected according to the reference scenarios defined in [6], where a distinction is made among three of them: urban, suburban and rural ones. All of them assume a homogeneous hexagonal deployment and uniform user distribution in order to carry out the different tests. Based on this premise, and taking into account those real networks constraints, the choice is to focus the work on two specific scenarios. In the first one of them, when the load decreases 3 out of 4 cells can be switched off, so just 1 remains active and a big amount of power is saved. This way, the Inter Site Distance (ISD) increases to the double of the size it had before the algorithm goes into action. Meanwhile, the alternative scheme presents the switching off of 8 out of 9 cells, increasing the ISD by a factor of 3. These two options can be seen in Fig. 1.

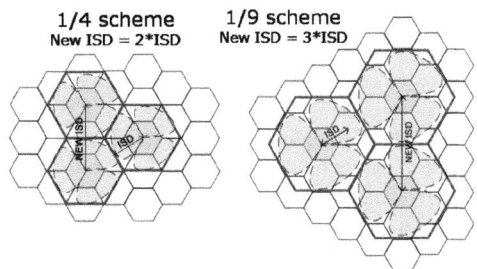

Fig. 1. 1/4 and 1/9 schemes

The deployment scenarios selected for the evaluation of the proposed solution derive from several simulations, extracted from [7] and used to help the authors to decide when and how to switch between different ISD scenarios (see Fig. 2, derived for simulations conducted within the EARTH project). Moreover, the SOTA model for power consumption, presented in Fig. 3, will be used to calculate the power transmitted in each case. By introducing instantaneous load of the BS (for instance, serving 10 Mbps when the maximum capacity is 50 Mbps will mean a 20% load), it is possible to directly obtain consumption in terms of Watts.

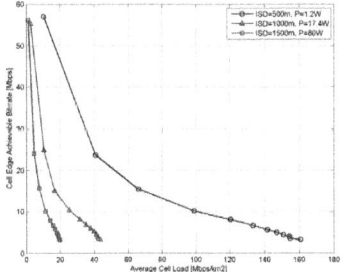

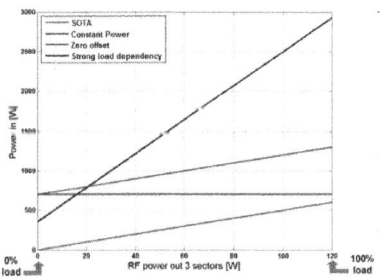

Fig. 2. Average cell load depending on the ISD and the Power transmitted

Fig. 3. Linear power models [7]

As regards reference system and main simulation parameters (described in [7]), a high traffic scenario (with peak traffic equal to 120 Mbps/Km^2) will be chosen as input, considering a macro BS with special capabilities to modify their transmitted power when changing the scenario (using MACRO BS able to reduce their power consumption when changing from 1/9 to 1/4 or 1/1 and not the other way around, because using MICRO BS enhancing their power would translate into big consumptions for 1/4 or 1/9 scenarios), and also taking in consideration 3 different ISDs (0.5, 1 and 1.5 Km) between cells that will derive in different cell load curves. In addition, a cubic interpolation of data will be using the power model for a Macro BS but making some corrections adapting it to tri-sectorial antennas, following the formula shown below.

$$P_{OUT} = 720 + 5 \cdot (3 \cdot P_{IN}) P_{OUT} = 720 + 5 \cdot (3 \cdot P_{IN}) \tag{1}$$

In this last formula, 720 factor corresponds to input power (due to physical components on the BS), and the 3 factor multiplying P_{IN} (introduced as % load) responds to
the tri-sector scheme.

2.1   Methodology and Data Retrieval from a Real Mobile Network

In order to study the performance of the ON-OFF scheme in a real network, the operation of the algorithm when it is fed with real traffic curves has been tested. In particular, a set of data traffic profiles was taken from Telecom Italia mobile network:
the dataset, extracted by a TIM mobile network monitoring system with a fixed time
resolution, is relative to an RBS currently running in the Telecom Italia mobile
network. In particular, for each sector, data samples automatically provided by the
monitoring system represent the average values of a specific KPI over 15 minutes
time intervals (e.g. for data bandwidth the sample represents the amount of data transferred in 15 minutes), and then a daily profile is represented by 96 consecutive values.
Each sample is also averaged over the 5 working days of the week, in order to
filter effects due to spurious data peaks and increase the reliability of the provided
profile.

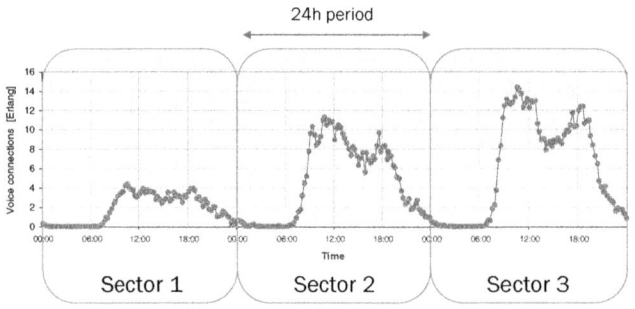

Fig. 4. Example of voice connections for UMTS sectors

Fig. 4 shows an example of the data profiles used, relative to a RBS installed and
running in the mobile network. The site considered to retrieve this data profile is a 3-
sectorial macro site located in an urban environment, equipped with GSM 900, GSM
1800, UMTS and HSDPA carriers, and captures all operator's traffic demand in the
covered area (2G, 3G and 3.5G). For easing the visualization, the 3 sectors are depicted in the same plot. The trend of extracted data profiles showed a typical daily
oscillation of the traffic, consistent with average profiles contained in [6][7], but with
the important difference that while literature curves are averaged over an entire network, data profiles used in this paper are relative to single sites, in order to better
show particular burst effects of the traffic variation. The resulting curves are then
suitable for a fine tuning of the developed prediction algorithms, and more reliable to
give a performance indication of the ON-OFF schemes in a real mobile network.

3 Prediction Methods

The appearance of several valleys during a regular day load imposes the introduction of additional mechanisms to improve the performance of the ON-OFF scheme. An important upgrade consists on combining instantaneous and historical data to take the most appropriate decision at any time. To do so, it is mandatory to gather the instantaneous data load together with a window of previous load conditions. The next step should be to combine this data with a predictor using coefficients historically tuned in order to determine the threshold of considering or not a short period.

Based on different daily load curves models, the evaluation started using diverse ON-OFF schemes: first static ones have been analyzed, and then the possibility of combining them dynamically has also been evaluated. So far, the difference lies in the number of ON-OFF schemes applied in the period of one day, this means from t = 0 to t = T = 24 hours; while the static model applies only one ON-OFF scheme a day, the dynamic model allows several transitions depending on the load. The upper side of Fig. 5 presents a comparison between power savings reduction using static and dynamic ON-OFF schemes applied on a simplified linear traffic model illustrated on the lower side of Fig. 5. In this case, two static ON-OFF schemes, 1/2 and 1/4 are compared with a dynamic model that combines both schemes. The static model only takes into account one specific threshold (either 1/2 or 1/4), whereas the dynamic model switches between both thresholds, applying the most appropriate in each situation. For the considered ON-OFF schemes, 1/x denotes that only 1 out of x base stations remain in ON state after the OFF management action.

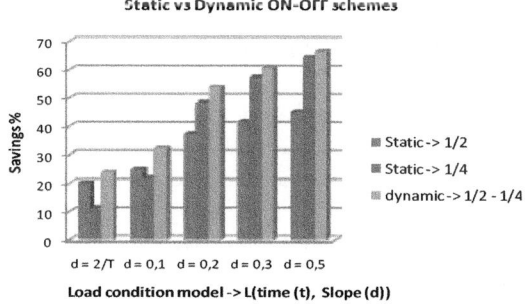

Fig. 5. Comparison between static and dynamic ON-OFF schemes in case of simplified linear traffic models

After seeing the theoretical energy savings in a simplified linear world, the next step is evaluating what potentials are in real networks [6][8]. Fig. 6 illustrates the energy saving potential in real life measurements provided by EARTH project partners. In the figure different static schemes are compared assuming a uniform omni-cell deployment where the daily variation of the traffic is according to real measurements. It can be concluded that more than 25% of the energy could be saved in an omni-cell deployment scenario.

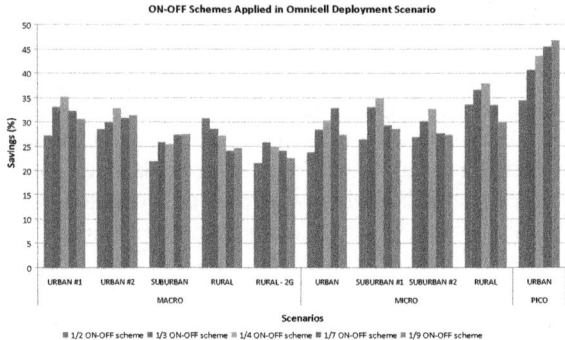

Fig. 6. ON-OFF schemes applied in omnicell deployment scenario

However, in contrast to the above linear model, the cell level traffic variation is substantially much more complex. Thus, other important aspect to be considered when applying the ON-OFF scheme is the necessity to limit the situations in which it is safe and helpful to apply this scheme. There might be situations in which several time slots in which it might be possible to apply an ON-OFF scheme exist. While some of them could actually be used, on others initiating the BS shut down might be counterproductive if the slot is not wide enough. For this purpose it is therefore essential to be aware of two critical issues: the time it takes for a BS to change state (from on to off or vice versa) and the expected value of the load on the cell during this time.

For the first one, a sensible assumption is to have slots larger than 30 minutes to have an efficient application of the algorithm, thus estimating conservatives on and off times. Pertaining to the network load on those future 30 minutes, some kind of predictor is needed in order to be able to foresee the future trends of the data curve. Therefore, in order to reach the pursued goals and avoid an inefficient use of the ON-OFF scheme, the utilization of some studies based on linear prediction will help. These kinds of predictors can tell whether the load curve is going to decrease (so the ON-OFF schemes can be applied) or it is going to increase after reaching some threshold, which would lead to a bad use of ON-OFF schemes. Initially the algorithms have been used mostly on voice call profiles (as data call profiles are very bursty), but trying to make a better approach the path taken has consisted in averaging real data curves. Such linear predictors are characterized by the following parameters:

- Model: Curve taken as reference to make predictions. In a system capable of learning, this model can change over time based on averages of the measures obtained daily. Similarly, several models that apply in different situations (weekdays, holidays...) can be used. In the case of ON-OFF algorithm, this model will be an averaged load curve. From here, an array of coefficients, which serve to weigh the actual data, is obtained.
- Window: Number of samples that are intended to predict from the actual data and model. The larger the window, the more errors will be committed, because we have to consider when making the prediction of all samples (except the first) entry must be entered as previous predictions, which are not accurate.

- Predictor order: Number of coefficients or measures taken as a reference from the model curve to construct the coefficient matrix. Higher orders cause predictions tightly adjusted to reality, but at the expense of increased processing times.

4   Power Savings

Applying all the previously mentioned inputs together (Averaged/Ideal data curve, High density urban scenario, BS power assumptions, linear predictors with 60th order and 30 minutes window), and evaluating a 24 hour period, it is possible to obtain theoretical savings, corresponding to ideal and smooth behaviour of data traffic. The primary savings are obtained when switching from 1/1 to 1/4 scheme. Additional ones are observed during night time, when it is possible to switch from this 1/4 scheme to 1/9. The total savings are 795.47 W/Km^2 in a period of 7 hours and 1 minute. It means that, on average, BS on our theoretical scenario, are wasting energy during 7 out of the 24 hours of the day.

Apart from the theoretical study of potential savings, tests in order to evaluate the predictor behaviour have been made. Using the linear predictor and making 1 million simulations evaluating data inputs based on the theoretical curves and then adding Gaussian noise with a dynamic variance and amplitude (random variance and maximum value for each simulation) of up to half of the maximum of the theoretical curve, the results were as follows:

- 9,101,495 predictions (steps for any of the thresholds)
- 8,138,433 correct predictions
- 963,062 wrong predictions

Doing the proper calculation, an 89.42% success rate is obtained using the predictors for this specific case of simulations with these windows, commands, data variability and thresholds. It is necessary to take into account many additional factors, such as time of day, the behaviour in previous days, the time of year, the number of current users in the cell,... All these features will be incorporated in future studies, but this first approach proves that the linear predictors are a very reliable alternative to combine with the ON-OFF algorithms in the real networks.

5   Conclusions and Further Work

The undertaken labour exhibited in this paper has shown the great promise that the selective schemes of switching on and off base stations offer. Fixing an optimal network configuration in each of the possible cases, it is relatively easy to decide when and how increase or reduce the power emitted by the BSs, thus achieving an interesting amount of savings. Really breakthrough achievements in the field of energy savings can be obtained using these solutions, tested over live networks load profiles.

Regarding further possible work, the introduction of predictors is a step forward in the complexity of the proposal, and thanks to them a new element of evaluation appears, which is critical when making decisions about whether the ideal time to undertake the switching on or off of a BS is approaching.

The present studies and the relative performance evaluations showed that the percentage of energy saved by applying these schemes is worthy of consideration, being necessary to validate such results in a real environment (This explains why the application of ON-OFF schemes to obtain gains in terms of power consumption in mobile networks has been selected by EARTH project as a track to be tested on a real infrastructure). To achieve the desired goals, the Telecom Italia Test Plant will be thus used in order to set up a test scenario with the necessary equipments (e.g. commercial eNB and terminals).

Acknowledgments. The work leading to this paper has received funding from the European Community's Seventh Framework Programme [FP7/2007-2013] under grant agreement n° 247733 – project EARTH. This publication made by participants of the EARTH project reflects only the author's views and the European Union is not liable for any use that may be made of the information.

References

1. Mendham, S.: The implementation of AMR metering at Vodafone UK, ESTA website (February 2008)
2. Vodafone, Vodafone Group announces commitment to reduce CO2 emissions by 50% (April 2008) (press release)
3. Marsan, M.A., Chiaraviglio, L., Ciullo, D., Meo, M.: Optimal Energy Savings in Cellular Access Networks. In: First International Workshop on Green Communications (Green-Comm 2009) (Dresden, Germany) (June 2009)
4. Oh, E., Krishnamachari, B., Liu, X., Niu, Z.: Toward Dynamic Energy-Efficient Operation of Cellular Network Infrastructure. IEEE Communications Magazine, 56–61 (2011)
5. Sánchez, L., Munoz, L.: Energy Efficiency of a Simple ON/OFF Scheme in Mobile Cellular Networks. Electronic Letters 46(20), 1404–1405 (2010)
6. EARTH project deliverable, D2.3, Energy Efficiency Analysis of the Reference Systems, Areas of Improvements and Target Breakdown (2010)
7. EARTH project deliverable, D3.1, Most Promising Tracks of Green Network Technologies (2010)
8. EARTH project deliverable, D2.2, Definition and Parameterization of Reference Systems and Scenarios (2010)

Challenge-Aware Traffic Protection
in Wireless Mobile Backhaul Networks

Javier Martín-Hernández[1], Christian Doerr[1],
Johannes Lessmann[2], and Marcus Schöller[2]

[1] TU Delft, Department of Telecommunication, Delft, Netherlands
{J.MartinHernandez,C.Doerr}@tudelft.nl
[2] NEC Europe Ltd., Heidelberg, Germany
{Johannes.Lessmann,Marcus.Schoeller}@neclab.eu

Abstract. To protect active traffic against link or node failures in multi-hop communications networks, several so-called protection schemes have been introduced in the past. The most established ones are path, segment, node and link protection. However, these schemes are limited as challenges are modelled abstractly whereas challenges in real networks can have very different characteristics. Thus, we propose to explicitly take the high impact challenges by introducing a risk-group concept into the multi-path placement scheme, which provides an evaluation of the likelihood of a challenge to simultaneously affect two network elements. We have implemented and evaluated this new methodology in simulations and show that it outperforms the original scheme.

Keywords: wireless mesh, wireless backhaul, challenge awareness.

1 Introduction

In the face of exploding user traffic in cellular networks, increasing the spectral efficiency to achieve higher access capacities is widely considered to be only possible via SDMA (space division multiple access). To this end, very large numbers of small cell base stations will be brought up in the near future. As an example, Picochip, a femtocell maker, claims that London needs to install 70,000 femtocells by 2015 to provide decent 4G LTE mobile services [1]. Since wired backhaul will not be generally available for all small cells, wireless backhaul networks will gain importance.

A drawback of a wireless backhaul network is the instability of its links. Even highly directed microwave links such as used for carrier-grade mobile backhaul networks are affected by bad weather conditions and link quality may degrade or a link might fail completely [2]. To avoid re-routing latency, a common approach is to proactively establish backup routes which are used in case the primary route fails. In the routing community, this is called multi-path routing, in the telecom world, this is known as protection. Given a source node, a destination node and a primary path between them, protection can happen at multiple levels of granularity. Figure 1 shows different widely established protection schemes.

Z. Becvar et al. (Eds.): NETWORKING 2012 Workshops, LNCS 7291, pp. 35–42, 2012.

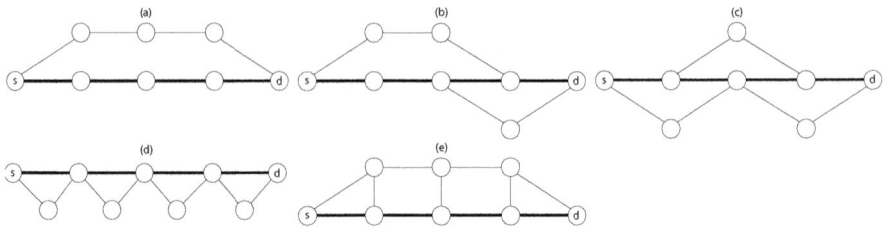

Fig. 1. Different protection schemes. a) path protection. b) segment protection. c) node protection. d) link protection. e) rope-ladder protection.

Rope-ladder protection (RLP) as introduced in [3] (cf. also Figure 1e)) combines the advantages of path, node and link protection by constructing two node-disjoint paths between s and d (i.e. "ropes") and connecting each node on the primary path with a node on the backup path (via "rungs"). As shown in [3], this increases path diversity and path lifetime while reducing loss gaps.

One of the major novelties of this paper over [3] is that it introduces the notion of *challenges* and thus challenge-awareness. A challenge to a wireless backhaul communication network could be a thunderstorm, a congestion hotspot or a virus attack, for example. Many protection schemes are either completely challenge-unaware or designed to meet only one particular challenge. In this paper, a path construction algorithm is proposed that can optimize the protection structure with respect to the high-impact challenges. Based on expert risk assessment, we apply the Shared-Risk Link Group concept known from the optical networks space [4] to the used protection scheme. A special entity of the network management system translates a challenge into a so-called *risk group*. Based on this, it steers the rope-ladder construction process such as to be maximally robust towards the high impact challenges. The mentioned special entity is the Graph Explorer, introduced in [5]. The Graph Explorer (GE) is a general tool that can explore a large set of properties in multi-hop networks. For the sake of this paper, we will use the capabilities of GE to compute the impact of challenges on the network and to steer the rope-ladder construction process accordingly. This will be described in detail in Section 3.

This paper is organized as follows. Section 2 discusses previous research in the domain of multipath QoS routing or protection and its relation to our proposition. Section 3 introduces the rope-ladder construction with the help of GE. Section 4 presents our performance evaluation. Section 5 gives a short summary.

2   Related Work

As mentioned previously, work in the context of our paper is discussed as multi-path routing as well as protection schemes.

There is quite a number of works in the domain of multipath routing protocols. An overview of this diverse field can be found in [6]. Different multi-path routing protocols focus on a number of aspects like load balancing, bandwidth bundling, security, congestion control or even security (sending packets of a sensitive flow via different paths to make eavesdropping more difficult). One goal is obviously resilience to failures. A further classification refers to the independence of the individual paths of the multipath. To increase security or robustness, the paths should be as independent as possible which gives rise to node-disjoint or link-disjoint multipaths. Some protocols do not make any statement or assumption about the path independence.

In the telecom world, particularly in the area of optical networks, related work is known as protection schemes. Terms like path, node or link protection are commonly used in traffic engineering technologies such as MPLS and do not need further discussion here. Interesting to mention, however, is that the optical networks community has introduced a concept called shared-risk link group (SRLG) [4]. A SRLG contains all links in a network that are susceptible to the same risk. The typical use case would be two optical fibers which share a common duct. If the duct is destroyed, likely not only one but both fibers share its fate, leading to disruption of traffic through both fibers. The concept of SRLG is very generic, allowing to capture arbitrary risks. Similarly, shared-risk node groups capture risks impacting one or multiple nodes. In [7], SRLG and SRNG are combined into shared-risk resource groups. Probably most of the SRLG related propositions occupy themselves with finding SRLG (SRNG/SRRG) diverse paths (path protection). However, they do not compare different protection schemes. In this paper, we will compare path and rope-ladder protection in the face of SRLGs.

3   Constructing Challenge Aware Protection Schemes

In this section, we describe how risk-aware protection schemes between a source and a destination node are constructed. First, we will introduce risk group models of three different challenges. Then we describe protection construction process based on the challenges which are to be considered.

3.1   Challenge Model and Risk Groups

A challenge is an event which occurs in the network and which threatens the network's normal operation. Examples for such challenges in wireless networks include for example adverse weather conditions, virus attacks, failures of software components, equipment theft or network overload.

As indicated previously, the optical networks domain has introduced the idea of shared-risk link groups (SRLG). Here, we apply this concept to multi-path protection. A challenge C is modelled in terms of such a risk group as a set of network elements failing simultaneously. The risk group of a challenge C, denoted as RG_C, is defined in terms of the *logical vicinity* of the protected elements (i.e. nodes and links): given that element e_1 is in risk group RG_C, element e_2 is also in RG_C if its logical vicinity to e_1 – denoted $v(RG_C, e_1, e_2)$ – is above a certain threshold τ. Formally, $n_1 \in RG_C \Rightarrow \forall n_2, v(RG_C, e_1, e_2) > \tau : e_2 \in RG_C$.

The impact of all challenges has been modelled with a step function (corresponds with setting $\tau = 0$) in our simulations. In other words, any link or node affected by the challenge C fails reducing its bandwidth to zero, hence $RG_C = C$. The impact function τ can easily be extended to complex and realistic scenarios. The logical vicinity function $v(RG_C, e_i, e_j)$ of all considered challenges needs to be defined manually by a network expert and can span one or more arbitrary dimensions. For instance, in areal challenges (e.g. a storm cell) the logical vicinity of two elements correlates with the geographical vicinity of the elements, whereas logical vicinity of a challenge exploiting a software bug correlates with the vendor ID.

In this paper, we have modelled three different classes of challenges: (i) a flash crowd event at a congestion hotspot, (ii) a heavy rain cell, and (iii) a virus targeting a firmware bug. A single **congestion hotspot** is defined by a static area, e.g., a train station or a stadium, where huge numbers of users can cause overload situations. All received connection requests are legitimate but cannot be satisfied by the system simultaneously; such events are called *flash crowd events* in contrast to denial of service attacks which are of malicious nature. The logical vicinity function is defined by the area the C affected elements are located at. The second condidered disruption is an areal **thunderstorm cell** moving randomly across the graph, producing a large set of independent thunderstorm challenges $\mathcal{C} = \{C_1, C_2...C_k\}$. The logical vicinity function of a single thunderstorm challenge C_i (for $i \leqslant k$) provides that elements e_1 and e_2 appear within the same risk group if a circular rain area with a radius $r(C_i)$ and an given epicentre $\varepsilon(C_i)$ overlaps both elements. Often, logical vicinity will be related to geographic positioning (e.g. distance to the epicentre) but other environmental characteristics may define this function, too. The last attack we considered is a generic virus attacking one **firmware vulnerability** of exposed graph elements, producing a set of firmware challenges $\mathcal{C}$. If a single firmware challenge C_i is launched against the mesh network, all nodes using the targeted firmware version $f(i)$ are threatened and hence share the same risk group.

3.2 The Construction Process

In order to construct challenge-aware rope-ladders, we have combined the RLP scheme with the Graph Explorer as introduced in [5]. The Graph Explorer is a tool to assess various metrics of a network in the face of an arbitrary number of challenges occurring simultaneously in this network. We use this tool to calculate the risk groups before starting to place the rope-ladder structure. The construction of a rope-ladder is divided in three sequential steps as follows.

Step A: Placement of the Primary Path. As the risk groups depend on the links and nodes of the primary path, the choice of the primary path is a crucial one in our process. An intuitive approach would choose the primary path to be the shortest path from source to destination. However this can lead to fragile backup paths crossing high risk elements. Thus, we propose that the process should iterate over all paths up to a maximum stretch with respect to the shortest path, providing a set $\mathcal{P}$ of paths. Eventually, the primary path which

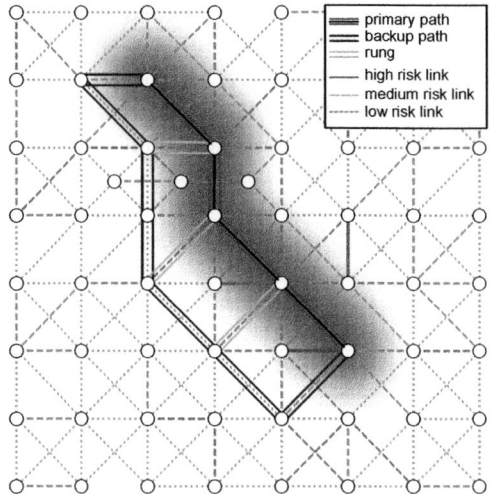

Fig. 2. A challenge-aware RLP scheme for one random Access Point and a Relay node on the G_B topology. The effect of the thunderstorm is displayed as a shadow covering the susceptible links, subject to the condition that the storm hitting the primary path. The darker the shadow, the higher the likelihood for a link to fail together with the primary path.

leads to the less risky backup path is selected as the primary path (as described in the last step).

Step B: Calculation of Link Weights. The input to this step are all the primary paths $\mathcal{P}$ provided by step A, and the high-impact set of challenges $\mathcal{C}$ from which the network should be protected. Depending on the chosen vicinity function, additional information must be made available such as firmware ID, the frequency allocation plan, etc. Multiple risk groups can be added into a unified risk group (URG) by merging the link weights of different challenge types. This merging function must be determined during the network manager's risk assessment process, and it should account for the respective occurrence probability of the different challenges types. In our simulations we assumed that all challenge types are both independent and equally probable. Hence all weights belonging to the same challenge type were further normalized to sum the complete probability. The output of step B are multiple *weight clouds* $\mathcal{W}$, i.e. sets of node and link weights representing the risk group memberships with respect to each primary path. This weight cloud calculation process is computed such that the weight of element e will increment for every time that e shared a challenge in $\mathcal{C}$ with any of the elements in $P \in \mathcal{P}$. An intuitive visual representation of each primary path's risk group is the union of all the challenge instances $\mathcal{C}$ (e.g. thunderstorms in this example) that intersect with the primary path by at least one link or node. The storm's link weights associated with the shortest path can be illustrated as the *cloud* shown in Figure 2.

Step C: Placement of the Backup Structure. A set of backup paths $\mathcal{B}$ is found by iterating over all URG pairs $\{P, W\}$ offered by step B. A shortest path algorithm determines the backup path with the least weight which does not exceed an arbitrary stretch limit. The basic idea of our approach is that the backup path circumvents the *cloud* in Figure 2 and stays out of it for as long as possible, hence minimizing the link weights that will be crossed. An enhancement to this step currently under investigation as part of our ongoing research is the use of a risk threshold heuristic which is acceptable for the backup path. The process would only iterate over the shortest paths until a backup path is found with acceptable risk, improving the overall time performance. Finally, once all $\{P, W\}$ pairs have been processed, the backup path with the lowest weight in $\mathcal{B}$ is selected to form the rope-ladder; the rungs are determined afterwards (the rungs are cross-connects from the primary path to the backup path to minimize the loss gap for real time application flows - see [3] for details).

4   Simulation Results

This section describes the simulation scenario built to evaluate the performance of challenge-aware RLP as introduced in Section 3, followed by a qualitative analysis of the simulation results.

4.1   Scenario Description

The selected application is a unicast VoIP application, simulating a G.711 VoIP codec over 1Mb/s duplex connections. To simulate the effect of a challenge on a voice stream, a voice call is held between two random nodes for an arbitrary time span of 3 minutes. This data flow is established via a primary path following a RLP scheme. One minute into the call, an instance of a challenge occurs; causing the bandwidth of the affected links to be reduced to zero for the duration of the challenge, virtually disconnecting them. As soon as the links become unavailable, the central routing protocol will divert in-flight packets and adapt routing tables to the RLP scheme backup path through the rung which is closest to the challenge. The challenge remains in place for one minute, after which all the links in the network are restored to their initial state.

4.2   Simulation Results

First, we measure the packet loss that different protection schemes suffer by a storm cell occurring. The percentage of packets lost by an oblivious RLP scheme (i.e. a rope ladder uninformed about possible challenges during construction) is 10.3%, approximately two times the percentage of packets lost by the RG_{Storm} aware RLP scheme, which lost 4.8% of packets as shown in the leftmost chart of Figure 3(a). Secondly we evaluate the gap size, measured as the the maximum difference in sequence numbers between two consecutive received frames. Given that the routing is controlled by a central authority, the delay induced

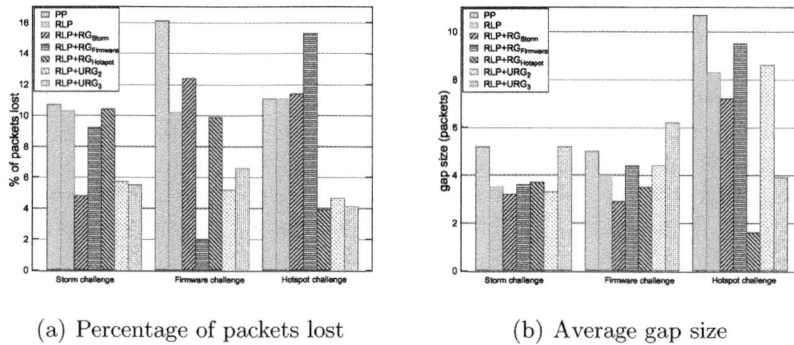

(a) Percentage of packets lost   (b) Average gap size

Fig. 3. (a) Percentage of packets lost and (b) average gap size of protection schemes. The six sets of charts display the gap size effects of *Storm* (left), *Firmware* (middle), and *Hostspot* (right) challenges on G_B. Each set displays seven bars, corresponding to the protection schemes: oblivious PP, oblivious RLP, RG_{Storm}-aware, $RG_{Firmware}$-aware, $RG_{Hotspot}$-aware, and URG_2-aware and URG_3-aware.

by a challenge message propagation is dismissed, i.e., the routing tables are instantly updated across the network. This effect works in favour of PP schemes, by ignoring the propagation delay involved in route table synchronization. Nevertheless the flows' gap size effect of rerouted in-flight packets is still noticeable. All three simulation scenarios displayed in Figure 3 illustrate RLP challenge aware schemes suffering smaller packet loss and gap sizes than PP oblivious schemes.

In order to test the behaviour of RLP schemes subject to challenges not included in the planned risk group RG_C we also measured the performance of a challenge aware RLP scheme against unexpected sets of challenges. The blue striped bars in Figure 3(b) illustrate this effect: challenge aware RLP schemes' performance degrades under the effect of unexpected challenges. The performance of schemes under unexpected attacks may even degrade beyond their oblivious counterparts. Such is the case for RLR RG_{Storm} schemes under Firmware challenges, which lost 4.4% of the voice packets, as opposed to only 4.0% for the oblivious scheme. This adverse effect motivated us to study multi-challenge aware protection schemes through the use of URG. Ideally a multi-challenge aware rope ladder structure can withstand different non-simultaneous attacks without a significant drop in performance. First we define URG_2 as the unified risk group formed by adding the two risk groups with the highest impact out of the storm cell, firmware virus and hotspot. Additionally we define URG_3 as the risk group resulting from the addition of all three risk groups. Weights are consequently normalized, as specified in section 3. Simulations show that the percentage of packets lost by URG_2 and URG_3 schemes under a storm cell challenge are 5.4% and 5.2% respectively (as illustrated Figure 3(a)), representing a significant improvement over the oblivious RLP scheme (10.3%), yet not performing as good as a RG_{Storm} aware scheme (4.8%). Overall, in terms of packet loss all URG protection schemes' outperform both their oblivious and challenge-aware schemes under the effects of unexpected challenges. On the other hand Figure 3(b) illustrates URG schemes suffering large gap sizes. The protection scheme with the lowest gap size is the one tailored to the challenge, i.e. $RG_{Hotspot}$ with a gap of 1.6 packets. However the gap sizes of URG_2 and URG_3 schemes under flash crowd challenges are 8.6 packets and 3.9 packets respectively, whereas the oblivious RLP scheme only lost 8.3 packets. In conclusion, URG schemes' gap sizes are highly dependant on the type of challenge and considered risk groups; these schemes may underperform their oblivious counterparts when faced to an expected sets of challenges.

5   Conclusion

In this paper, we have presented an algorithm to improve the placement of rope-ladder protection schemes in multi-hop wireless networks. The algorithm is based on the Graph Explorer, a general tool to explore properties and metrics in arbitrary graphs. During the network planing phase, we have employed a risk group approach which makes use of a logical vicinity function that relates each link and node in the network to individual risk groups. During network operation the Graph Explorer assesses possible placements of rope-ladders such as to be maximally robust towards certain challenges. Using simulations, we have evaluated this challenge aware rope-ladder scheme with the original rope-ladder scheme and with path protection. The packet loss rate was reduced by up to 80.4% compared to the oblivious scheme, but more remarkably the number of protection schemes surviving the challenge onset was increased by up to 25%. Focusing on the high impact challenges during the network design stage is critical.

Acknowledgements. The research leading to these results has been funded by the European Commission, under grant agreement no. 224619 (ResumeNet).

References

1. http://eweekeurope.co.uk/news/
   london-needs-70000-cells-for-4g-broadband-40779
2. Jabbar, A., Raman, B., Frost, V.S., Sterbenz, J.P.G.: Weather Disruption-Tolerant Self-Optimising Millimeter Mesh Networks. In: Hummel, K.A., Sterbenz, J.P.G. (eds.) IWSOS 2008. LNCS, vol. 5343, pp. 242–255. Springer, Heidelberg (2008)
3. Lessmann, J., Schöller, M., Zdarsky, F., Banchs, A.: In: Proceedings of the 2010 IEEE WoWMoM, WOWMOM 2010. IEEE Computer Society (2010)
4. Sebos, P., Yates, J., Hjalmtysson, G., Greenberg, A.: In: Optical Fiber Communication Conference, WDD3. Optical Society of America (2001)
5. Doerr, C., Hernandez, J.M.: In: Third International Conference on Dependability, DEPEND (2010)
6. Mueller, S., Tsang, R.P., Ghosal, D.: Multipath Routing in Mobile Ad Hoc Networks: Issues and Challenges. In: Calzarossa, M.C., Gelenbe, E. (eds.) MASCOTS 2003. LNCS, vol. 2965, pp. 209–234. Springer, Heidelberg (2004)
7. Datta, P., Somani, A.K.: Comput. Netw. 52, 2381–2394 (August 2008)

A Distributed Signaling Fast Mobile IPv6 Scheme for Next Generation Heterogeneous IP Networks

Mohtasim Abbassi, Shahbaz Khan, and M. Rahman

Department of Telecommunication Engineering,
University of Engineering & Technology,
Peshawar (Mardan Campus), 23200, Pakistan
mohtasimabbassi@yahoo.com

Abstract. The Next-Generation Wireless Networks (NGWNs) are believed to exhibit a heterogeneous environment which imposes various challenges, among them Mobility Management is most crucial. The FMIPv6 protocol has been standardized by IETF to address certain shortcomings of the baseline Mobile IPv6. However, it does not directly support vertical handovers. Therefore, research is still going on to develop more effective IPv6 based vertical handoff protocols for efficient support of frequent and seamless roaming. This paper presents a new protocol named Distributed Signaling Fast Mobile IPv6 (DS-FMIPv6) which effectively addresses reduction of handoff latency, signaling costs and power overheads for a better support of seamless & frequent vertical handoffs.

Keywords: Mobility Management, NGWNs, FMIPv6, DS-FMIPv6.

1 Introduction

The Next-Generation wireless networks are extensively recommended to be all-IP-based. Therefore, the IPv6-based Mobility Management protocols are of significant importance. The FMIPv6 [1] proposed by IETF aims towards improving the performance of the fundamental MIPv6 protocol [2]. These protocols are however hindered by several drawbacks of which the handoff latency is most important. Moreover, in case of frequent handovers, signaling and power overheads incurred by these schemes become even more critical. These protocols also require improvements to become suitable enough to efficiently fulfill several stringent requirements like context transfers during vertical handoffs.

In this paper, a vertical handoff protocol named Distributed Signaling FMIPv6 (DS-FMIPv6) is presented. The signaling involved in this handoff protocol is distributed among MN, Current Access Router (CAR) and the New Access Router (NAR) - hence the scheme is named as Distributed Signaling FMIPv6 scheme. The performance of FMIPv6 is further improved in DS-FMIPv6 by carrying out selected signaling procedures in advance i.e. prior to the start of actual handover and by increasing the functionality of routers. The protocol operation relies on a network selection algorithm evaluated at the Current Access Router (CAR) which evaluates the best

Z. Becvar et al. (Eds.): NETWORKING 2012 Workshops, LNCS 7291, pp. 43–51, 2012.

suitable network beforehand for the current MN status. This algorithm is assumed to encompass all necessary parameters required for network selection. Detailed analysis shows that DS-FMIPv6 scheme significantly reduces the handoff latency along with signaling and power overheads when an MN undergoes frequent handovers.

The rest of paper is organized as follows: in section 2, an overview of IETF-proposed FMIPv6 protocol along with its few extensions is presented. Section 3 presents the proposed DS-FMIPv6 protocol. Performance analysis of DS-FMIPv6 is explained in Section 4. Finally, section 5 concludes the paper and sets directions for future research in connection with this topic.

2   Fast Mobile IPv6 and Its Extensions

2.1   Fast Mobile IPv6

FMIPv6 is proposed to reduce handoff delay and to minimize the services disruption incurred with MIPv6 during handoffs. The most invigorating feature of FMIPv6 is that it relies on link layer information (L2 Trigger) to predict or to rapidly respond to handoff events (Predictive and Reactive mode respectively). When an MN detects its movement towards NAR, by using L2 Trigger, it exchanges Router Solicitation for Proxy Advertisement (RtSolPr) and Proxy Router Advertisement (PrRtAdv) messages with Previous Access Router (PAR) to configure a new Care-of Address (nCoA). The MN sends FBU to PAR in order to associate its previous CoA with New CoA. In order to prevent routing failure, a bidirectional tunnel between PAR and NAR is established by using Handover Initiate (HI) and Handover Acknowledgment (HAck) messages. On communicating FBU to PAR, the MN undergoes L2 handoff procedures. After successful L2 Handoff the MN announces its presence on new link by using Unsolicited Neighbour Advertisement (UNA) message. The NAR at this stage starts delivering packets to the MN. The Binding Updates to HA & CN are carried out in similar fashion as for MIPv6.

2.2   Enhancements in Fast Mobile IPv6

Various enhancements in FMIPv6 are proposed with different aims and objectives, but their efficiency is hindered by certain limitations. An overview of few schemes is presented in this sub-section. In [3], the handoff latency of FMIPv6 is improved by making certain fundamental enhancements in various processes involved. The movement detection & tunnel establishment procedures are merged, resulting in reduction of latency because of DAD procedure. The Binding Update to the HA/CN is brought forward to the time before the layer 2 handover which further improves the latency problem. In [4], an Intelligent FMIPv6 (iFMIPv6) scheme is designed specifically for WLANs. This scheme addresses the reduction in Scanning Delay, where the Access Point will have Pre-knowledge of its neighbourhood and the MN is intelligent enough knowing its mobility pattern and thus the next stop of its movement. Therefore, at the time of handoff, the MN only needs to scan the selected AP by probing the corresponding operating channel, thus reducing the overall scanning delay. In [5], an efficient handover scheme between WLAN and 3G-UMTS is proposed utilizing Pre-Authentication in order to achieve a lossless vertical handover. An efficient handoff decision algorithm provides three link layer triggers for setting off the MN to detect the movement to a new subnet, to start Pre-Authentication, and to initiate Fast Binding Update. This Pre-Authentication concept reduces the overall service disruption time to a reasonable level. In [6], FMIPv6 based cross layer optimized vertical handover scheme is proposed for Mobile WiMAX, UMTS and CDMA networks. In this scheme, certain Layer 2 and Layer 3 messages are reordered or combined to eliminate redundant signaling. In [7], pseudo-binding FMIPv6 (pFMIPv6) scheme is presented which addresses the fast moving issue of the MN. The Access Router informs mobile node about the existence of other networks around the current network. The MN in turn creates several CoAs and initiates pseudo Fast Binding Update (pFBU) with all its neighbouring networks, even though it does not detect the signal of the next network yet. In this way, time deficiency involving these HO procedures for a fast moving MN is reduced to a high extent.

3   The DS-FMIPv6 Protocol

The signaling involved in DS-FMIPv6 is majorly derived from the MIPv6 and FMIPv6. The FBAck message is omitted and a new message Binding Update Initiate (BUI) is defined for its efficient protocol operation. Following steps constitute the protocol operation of DS-FMIPv6

i. Any change in status of MN (e.g. its average velocity, direction or imminent active application) is reported to the CAR through an enhanced RtSolPr (eRtSolPr) message. The RtSolPr message in FMIPv6 is enhanced to form eRtSolPr by defining a new option field which carries information about the above defined parameters. The Link Local Address (LLA) of MN is also communicated using LLA option.

ii. The CAR evaluates the handoff decision algorithm (HDA) using MN's information from eRtSolPr message and neighbouring networks information from the Router's cache. The prospective new network of MN is evaluated and new Care-of Address (nCoA) is formulated using LLA and network prefix. [8]

iii. The CAR now initiates the Duplicate Address Detection (DAD) to test the uniqueness of the CoA, by exchanging pHI (pseudo HI) and pHAck (pseudo HAck) messages with NAR. This exchange of messages also establishes a tunnel between CAR and NAR, which initially remains inactivated. The establishment of tunnel beforehand can be more effective in reduction of packet loss during handovers.

iv. CAR sends ePrRtAdv (enhanced PrRtAdv) message to inform MN its most suitable network by communicating its new CoA to it.

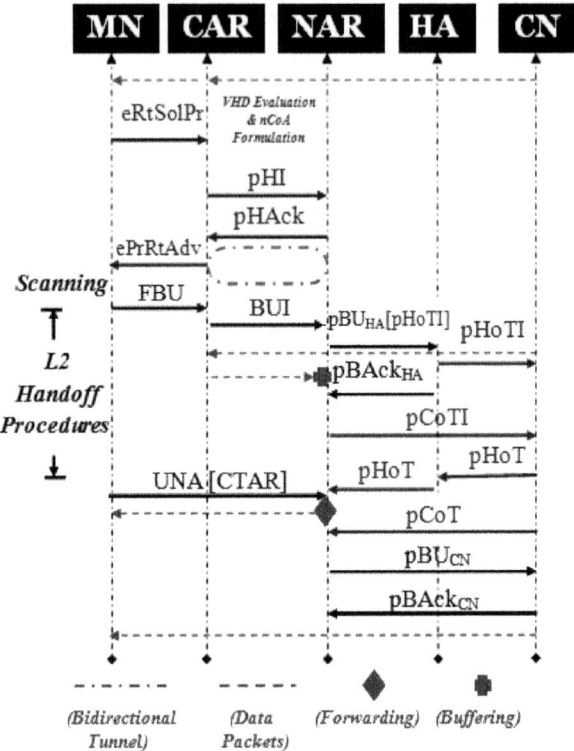

Fig. 1. Signaling Sequence of DS-FMIPv6 Scheme

v.   Acquiring nCoA, the MN sends FBU message to inform CAR that it is ready to handoff to new network. The CAR thus discontinues forwarding traffic to it and starts forwarding packets to NAR through already established tunnel. The CAR also sends a Binding Update Initiate (BUI) message to NAR piggybacking CTD (Context Transfer Data) [9] message in order to inform it that a new MN is about to join it, and instructs it to initiate the Binding Update procedure with CN/HA.

vi. The MN, on the other hand, scans new network and initiates Layer 2 (L2) handoff. It then announces its presence on the NAR through UNA message. CTAR (Context Transfer Activate Request) message is also piggybacked on it, which carries an authorization token to ascertain the reliability of the Context Transfer. At this instant the protocol operation involving MN completes and MN is ready to receive data packets from NAR routed through CAR.

vii. The Binding Update procedure on NAR begins when the pHoTI message piggybacked on pBU_{HA} (pseudo-BU_{HA}) is sent to HA. This allows the simultaneous initiation of pBU procedure to HA and Return Routability (RR). The HA then forwards pHoTI to CN(s), and responds the NAR by pBA_{HA} (pseudo-BA_{HA}) message.

viii. The NAR also sends the pCoTI (pseudo CoTI) message to the CN. The response of pHoTI and pCoTI message from CN is pHoT and pCoT messages respectively. The pHoT message is routed to NAR via HA. The RR procedure completes at this stage.

ix. After completion of RR, the NAR can now send the pseudo-Binding Update to Correspondent Node (pBU$_{CN}$). The CN in turn responds with a pBA$_{CN}$ (pseudo-BU$_{CN}$) message to NAR. At this instant the NAR will start receiving data traffic directly from CN(s). This completes the DS-FMIPv6 protocol operation. The signaling flow between different entities is shown in the Figure 1.

4   Performance Analysis

4.1   Handoff Latency

The Handoff Latency is referred to as the time interval from the moment that packets cannot be sent or received to the moment that the mobile node can communicate directly with CN via new Access Router [3]. The overall handoff delay for FMIPv6 and DS-FMIPv6 are given as;

$$T_{FMIPv6} = T_{DAD} + T_{FBAck} + T_{L2} + T_{UNA} + T_{BU}$$

$$T_{DS-FMIPv6} = \max\{T_{L2} + T_{UNA} + T_{arrival}, T_{BUI} + T_{pBU}\}$$

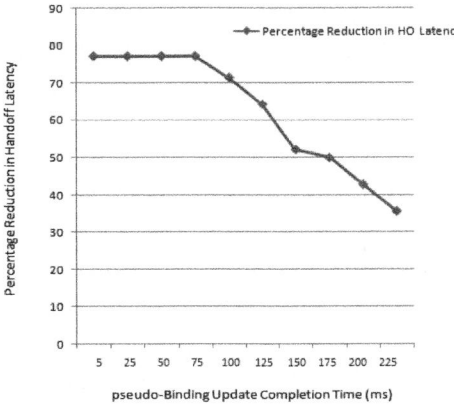

Fig. 2. Percentage Reduction of Handoff Latency in DS-FMIPv6 in comparison to FMIPv6

4.2   Signaling Cost

The Advanced DAD procedure [10] incorporated in DS-FMIPv6 eliminates the probability of any Reactive Mode. Here for comparison, we will assume successful anticipation of handover and will compare the signaling of predictive mode of FMIPv6 with DS-FMIPv6.

48   M. Abbassi, S. Khan, and M. Rahman

$$C_{FMIPv6} = 4C_{MN,PAR} + 3C_{PAR,NAR} + 2C_{MN,NAR} + 5PC_{AR} + C_{BU} + C_{RR}$$

$$C_{DS-FMIPv6} = 3C_{MN,CAR} + 4C_{CAR,NAR} + 2C_{MN,NAR} + 6PC_{AR} + C_{pBU}$$

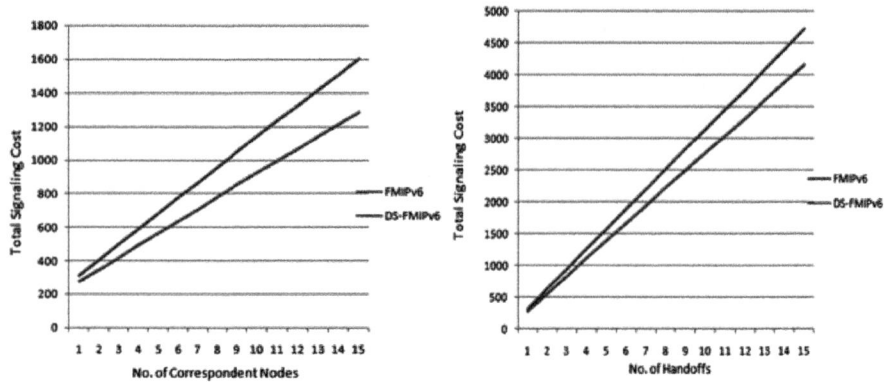

Fig. 3. Signaling Cost comparison of FMIPv6 and DS-FMIPv6

4.3   Power Consumption

Evaluation of a supportive handoff decision algorithm at Access Router in DS-FMIPv6 facilitates MN to turn on only the suitable network interface at the time of handoff. Traditionally, there can be three other possibilities for interface activation of MN i.e. all interfaces active all the time, active periodically, or at the time of handoff. The power consumption for each of these cases denoted by P_D, P_A, P_P and P_H respectively is mathematically formulated as;

$$P_A = P_S + \sum_{n_h} t_s \cdot \sum_{N-1} P_{IDLE} + n_h \cdot t_{HO} \cdot \sum_1^N P_{HO}$$

$$P_P = P_S + \frac{0.5}{3}\left(\sum_{n_h} t_s \cdot \sum_{N-1} P_{IDLE}\right) + n \cdot t_{HO} \cdot \sum_N P_{HO}$$

$$P_H = P_S + n_h \cdot t_{HO} \cdot \sum_N P_{HO}$$

$$P_D = P_S + n_h \cdot t_{HO} \cdot P_{HO}$$

For comparison, we consider an MN undergoing 15 handoffs among four networks with different power consumptions in each state [11, 12]. Results show that interface management in DS-FMIPv6 incurs the lowest possible power consumption.

Table 1. Example Scenario for Power Consumption comparison

Interface State	ID	UL	DL	UL	ID	UL	DL	DL
Network	GSM	WLAN	UMTS	GSM	UMTS	WLAN	CDMA	WLAN
$t_s(s)$	45	20	70	20	35	120	80	90
Interface State	UL	ID	DL	DL	UL	DL	ID	DL
Network	UMTS	WLAN	GSM	CDMA	UMTS	WLAN	GSM	CDMA
$t_s(s)$	60	30	140	80	40	100	30	40

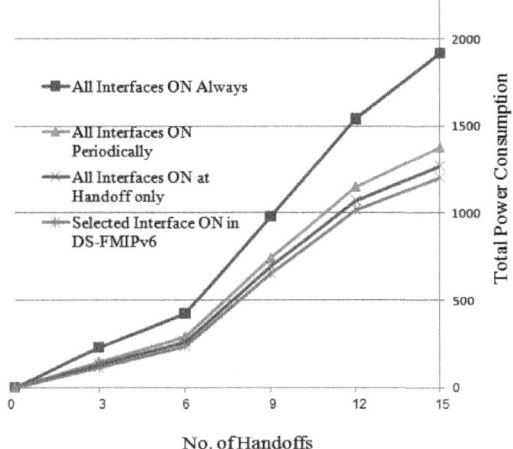

Fig. 4. Power Reduction by DS-FMIPv6 Interface Management

Table 2. System Parameters

Parameter	Notation	Value
(a). Handoff Latency		
Layer 2 Handoff Delay	T_{L2}	50 ms
Message Processing Delay at any Node X	$T_X(X)$	5 ms
Link Delay (among Local Nodes A & B)	$T_X(A, B)$	10 ms
Link Delay (among Remote Nodes like HA/CN)	$T_X(A, HA/CN)$	15 ms
(b). Signaling Cost		
Signaling Cost between MN & PAR/CAR	$C_{MN, PAR/CAR}$	5
Signaling Cost between PAR/CAR & NAR	$C_{PAR/CAR, NAR}$	7
Processing Cost at any AR	PC_{xAR}	8
Signaling Cost between MN & HA	$C_{MN, HA}$	15
Signaling Cost between any AR & HA	$C_{xAR, HA}$	10
Signaling Cost between MN & CN	$C_{MN, CN}$	15
Processing Cost at HA	PC_{HA}	24
Processing Cost at CN	PC_{CN}	4
Signaling Cost between HA & CN	$C_{HA, CN}$	10
(c). Power Consumption		
Power consumption per second in State "x"	P_x	*Manifold (Table 3)*
Session time for MN in State "x"	t_x	*Manifold (Table 3)*
Interface Activation time for Handoff	t_{HO}	1 sec

Table 3. Partial Expressions

(a). Handoff Latency
$T_{DAD} = T_{HI}(PAR) + T_{HI}(PAR, NAR) + T_{HI}(NAR) + T_{HAck}(NAR, PAR) + T_{HAck}(PAR)$
$T_{xBU} = T_{xBU_{HA}} + T_{xBU_{CN}} + T_{xRR}$
(b). Signaling Cost:
$C_{BU} = 2\left(C_{MN,HA} + N_{CN} \cdot C_{MN,CN}\right) + PC_{HA} + N_{CN} \cdot PC_{CN}$
$C_{RR} = 2\left(C_{MN,HA} + N_{CN} \cdot C_{HA,CN} + N_{CN} \cdot C_{MN,CN} + PC_{HA} + N_{CN} \cdot PC_{CN}\right)$
$C_{DBU} = 4\left(C_{NAR,HA} + N_{CN} \cdot \left(C_{NAR,CN} + C_{HA,CN}\right)\right) + 3\left(PC_{HA} + N_{CN} \cdot PC_{CN}\right)$
(c). Power Overhead:
$P_S = t_{DL} \cdot P_{DL} + t_{UL} \cdot P_{UL} + t_{IDLE} \cdot P_{IDLE}$
$t_s = t_{DL} + t_{UL} + t_{IDLE}$

5   Conclusion

The IETF solutions proposed for Mobility Management are not compact enough to support vertical handoffs. Also their handoff delays adversely affect delay sensitive and real-time applications. Other issues of signaling Cost and Power consumption at MNs become equally crucial in case of recurring handoffs. In this paper, a novel DS-FMIPv6 protocol is proposed which effectively reduces the handoff latency imposed by FMIPv6 from 35-75%. The proposed scheme also promises the lowest possible signaling cost & power consumption of MNs. In future, the obtained numerical results will be validated through simulations by modeling the protocol on a suitable simulator in order to provide a better analysis of these metrics.

References

1. Koodli, G.: Fast Handover for Mobile IPv6, IETF RFC 5568
2. Johnson, D.B., Perkins, C.E., Arkka, J.: Mobility Support in IPv6, IETF RFC 3775
3. Li, R., Li, J., Wu, K., Xiao, Y., Xie, J.: An Enhanced Fast Handover with low latency for Mobile IPv6. IEEE Transactions on Wireless Communications 7(1) (January 2008)
4. Zhang, L.J., Zhang, L., Pierre, S.: Intelligent Fast Handover Scheme for Mobile IPv6-based Wireless Local Area Networks. IJCSNS 9(8) (August 2009)
5. Menezes, S., Venkatesan, S., Rho, K.H.: An Efficient Handover Scheme based on Fast Mobile IPv6. In: IEEE 64th VTC-2006 Fall, pp. 1–5 (September 2006)
6. Jo, J., Cho, J.: A Cross-Layer Vertical Handover between Mobile WiMAX and 3G Networks. In: IWCMC 2008, pp. 644–649 (August 2008)
7. Kim, H.: An Enhancement of FMIPv6 for Packet Radio Networks which Supports the QoS Provisioning on the MIPv6. JDCTA 3(2) (June 2009)
8. Thomson, S., Narten, T., Jinemei, T.: IPv6 Stateless Address Autoconfiguration. IETF RFC 4862 (September 2007)
9. Loughney, J. (ed.), Nakhjiri, M., Perkins, C., Koodli, R.: Context Transfer Protocol (CXTP). IETF RFC 4067 (July 2005)

10. T-Trong, S., Tursunova, S., Kim, Y.-T.: Enhanced Vertical Handover in Mobile IPv6 with Media Independent Handover Services and Advance Duplicate Address Detection. In: KNOM Conference (2008)
11. Salawu, N., Onwuka, E.N.: Energy Optimization Mechanism for Mobile Terminals using Vertical Handoff between WLAN and CDMA2000 Networks. Leonardo Electronic Journal and Practices and Technologies (15), 51–58 (July-December 2009)
12. Perrucci, G.P., Fitzek, F.H.P., Sasso, G., Kellerer, W., Widmer, J.: On the Impact of 2G and 3G Network Usage for Mobile Phones' Battery Life. In: EW 2009, pp. 255–259 (December 2009)

Energy Efficiency Gains Using VHOs
in Heterogeneous Networks

António Serrador and Luís M. Correia

Instituto Superior Técnico/Instituto de Telecomunicações,
Technical University of Lisbon, Portugal R. Rovisco Pais, 1049-001 Lisbon
{ajns,luis.correia}@lx.it.pt

Abstract. Partially, the future of mobile and cellular networks will be about energy consumption savings, at the network infrastructure side. In the heterogeneous cellular networks environment, these can be achieved based on bandwidth adaptation, discontinuous transmission, relays, or even Base Stations (BSs) shutdown when possible. This paper proposes an approach for power savings using Vertical Handovers (VHOs) among different Radio Access Technologies (RATs), based on Radio Resource Management. The presented solution use a utility function focused on the energy efficiency of a given BS and RAT. An active VHO trigger mechanism is also proposed to provide RF power savings by handover mobile terminals to the most efficient RAT. Based on a power model proposed by the ICT-EARTH project, it is possible to translate RF power savings into system level ones. Results show that VHOs RF power gains range up to 46%, and up to 13% at the BS system level.

Keywords: Energy Efficiency, Radio Resource Management, Vertical Handover, Heterogeneous Networks.

1   Introduction

In wireless mobile heterogeneous network environments, Radio Resource Management (RRM) and Joint RRM (JRRM) entities, and their corresponding functionalities/algorithms, can be used to manage modern networks focused on different goals. One of them is definitely networks infrastructures energy consumption, and the corresponding impact on networks operators cost, such the Operational Expenditure (OPEX), simultaneously addressing the carbon dioxide emissions problem. Nowadays, the whole Information and Communication Technologies (ICT) industry [1]-[2] is responsible for about 2% of CO_2 emissions worldwide, energy cost ranging from 20 to 35% of operators OPEX. Thus, it is very important to decrease power consumptions in all areas, using imaginative techniques.

In recent literature and research projects, Vertical Handovers (VHOs) are used to save energy in Mobile Terminals (MTs), especially the new 4G ones. In [3], authors use VHOs between IEEE 802.11 and 802.16e systems to reduce terminals power consumption. Detailed power and performance models for both standards are defined to evaluate the VHO opportunity. More recently, European projects are dedicated to

Z. Becvar et al. (Eds.): NETWORKING 2012 Workshops, LNCS 7291, pp. 52–62, 2012.
© IFIP International Federation for Information Processing 2012

Energy Efficiency (EE) in mobile networks: the EARTH project [1] is devoted to reduce networks infrastructure power consumption, using a wide set of saving techniques; C2POWER [4], addresses EE optimisation in MTs, assuming multi-Radio Access Technologies (RATs) capabilities, attaching MTs to the most efficient radio link. Authors in [5] also address VHOs as a fundamental technique, by proposing a RAT selection algorithm that selects a RAT with lower energy consumption to increase terminal battery life, while respecting QoS. In [6], authors propose EE VHOs by reducing MTs frame overhead. Nevertheless, relevant network infrastructure EE using VHOs is not addressed.

In this work, VHOs are used to provide additional EE gains in the network infrastructure and not only at MTs. The idea is as follows: in a multi-RAT environment, legacy RATs are usually less energy efficient when comparing power and information processing, thus, moving active MTs (and corresponding services) to modern RATs provides additional EE, assuring a reasonable QoS balance among RATs. Aiming at this goal, implies taking three steps: the EE metric is identified and used in a Cost Function (CF); a VHO trigger algorithm is implemented and proposed; a system level power model is computed to obtain the overall energy gain in a given Base Station (BS), knowing the radio power gain.

This approach assumes the existence of a JRRM entity in a very tight coupling architecture, being simulated by a multi-RAT simulator tool [7], capable of simulating radio interfaces such as Wideband Code Division Access (WCDMA) and High Speed Downlink Packet Access (HSDPA) defined by the 3rd Generation Partnership Project (3GPP); also, traffic source models are implemented, such as voice, web browsing and File Transfer Protocol (FTP). By using this tool, micro-cells can be simulated in an urban dense scenario.

The main goal and novelty of this paper is to obtain the EE gains at radio and BS system levels, by exploring EE margins in a multi-RAT environment, using VHOs and relevant metrics to trigger handovers.

The paper is structured as follows. Section 2 presents the BSs power blocks and the power relation between RF and system levels. In Section 3, the used metrics, CFs and VHOs algorithm are proposed, while in Section 4, the fundamental reference scenario is presented for micro-cells and corresponding traffic loads, computational results being discussed. Finally, conclusions are drawn in Section 5.

2   BS Power Model

This section defines the BS micro-cell power model consumption defined by the EARTH project 8. It is assumed that the 2010 BSs systems state of the art power conditions are available. This model is relevant because it enables the computation of system level power gains, knowing the RF EE gain provided by the use of VHOs or another RRM technique. Thus, the model described in this section is used to estimate the BS complete system power gains, knowing the RF power savings provided by techniques or algorithms running at the air interface.

The power consumption of BSs is dominated by the radio supporting equipment. In general macro-, micro-, pico- and femto-cells BS equipment blocks include multiple transceivers (TRXs) with multiple antennas. Each of these TRXs comprises a loss Antenna Interface (AI), a Power Amplifier (PA), a Radio Frequency (RF) small-signal transceiver section and a baseband interface (BB) including both a receiver (UL) and a transmitter (DL) section, a DC-DC power supply regulation, an active cooling system, and finally a main AC-DC power supply for connection to the electrical power grid. Note that active cooling is only relevant for a macro-cell BS, being negligible for smaller cells. Details about this model are provided in 8.

In Fig. 1, the BS overall power consumption dependency on RF output power is highlighted.

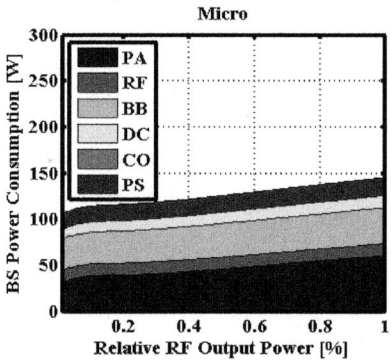

Fig. 1. Micro-cell RF and BS power relation (extracted from [8])

The EARTH project proposes a simple relation between RF and BS power as follows:

$$P_{BS} = P_0 + \Delta_P \cdot P_{RF} , \quad 0 \leq P_{RF} \leq P_{max} \tag{1}$$

where,

- P_{BS}, is the overall required BS power;
- P_{RF}, is the output RF power;
- P_o, is the BS power consumption calculated at the minimum possible RF power;
- Δ_P, is the linear P_{BS} relation with P_{RF}, for each BS type;
- P_{max}, is the maximum RF power.

3   JRRM and RRM Models

3.1   EE Metrics and Cost Function

In general, transmission of user data is accompanied by additional network signalling. In some cases, even in an "empty" network, radio beacons transporting network information must be continuously transmitting useful data (e.g., BS identification) to standby MTs. Moreover, user data includes other information, such as, radio channel formatting and coding overhead, high network layers protocol additional information, and retransmission of information when radio channels are of poor quality. Therefore, in the end, the relation between real user data and overhead is relevant when discussing EE, and the relation between transmitted bit and power. One can divide the previous concepts into the following parameters, Fig. 2:

- The information volume related to overhead and signalling (includes network and users signalling data) issues, as well as retransmissions, V_O;
- The information volume related only to user real data generated by the application, V_D;
- The total information volume V_T, required to transmit/support the user data, is V_O plus V_D.
- The power required to transmit the overhead information V_O, including network and users signalling data, is P_O;
- The power required to transmit user information V_D is defined as P_D;
- The total power required to transmit V_T is assumed to be P_T.

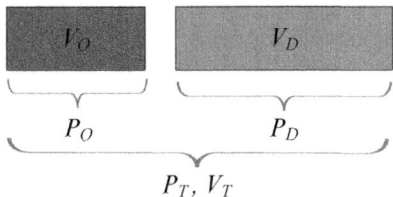

Fig. 2. Network and users' power and information parameters

Additionally, the BS energy cost can be established, being assumed as an EE metric of a given BS and associated RAT. Thus, the total power cost required to transmit V_D, is assumed to be C_U, given by:

$$C_{U[\text{W/bit}]} = \frac{P_{T[\text{W}]}}{V_{D[\text{bit}]}} \tag{2}$$

This power cost C_U is a key metric to measure the EE of a given BS; however, appropriate RRM decisions should be based on an opportunistic time scale, being crucial to evaluate BSs efficiency, therefore, C_U can also be measure in W/bps. This metric is computed by a CF defined in 9 and 10, being used in this work only with the operator component. Different CFs may be computed for each different RAT type. Each one of these "sub CFs" are weighted with different values, enabling the implementation and evaluation of different policies by RRM algorithms over each type of RAT. The CF model can be used at different network levels as follows: network KPIs (Key Performance Indicators), MTs, BSs, RATs, and operators. In order to combine all these different parameters, they must be normalised. For operators, the total cost for a given RAT r, Co_r, is calculated as follows:

$$Co_r = \frac{1}{N_{BS_r}} \sum_{b=1}^{N_{BS_r}} Co_{r,b} \tag{3}$$

where,

- N_{BSr} is the total number of BSs for a given RAT r,
- $Co_{r,b}$ is the operator's cost for each BS b, in RAT r, being computed by:

$$Co_{r,b} = \frac{1}{\displaystyle\sum_{i=1}^{N_{KPI_r}} w_{r,i}} \sum_{i=1}^{N_{KPI_r}} w_{r,i} \cdot k_{b,i} \qquad (4)$$

where,

- N_{KPI_r} is the total number of KPIs of a given RAT r;
- $w_{r,i}$ is the weight of each KPI i;
- $k_{b,i}$ represents the normalised value of each KPI ($0 \le k_{b,i} \le 1$).

Note that KPIs can be any relevant parameter that is useful for the management of cellular networks. In this paper, the power cost C_U metric is used as a KPI. Modern RRM algorithms can use this CF model or principles to guide networks focused on their EE metrics (among others), providing fast methods to distinguish the most efficient BSs and RATs at different networks operation time scales.

3.2   RRM Algorithms

In 11, an algorithm to trigger VHOs is proposed, the so called Fittingness Factor (FF). This algorithm is divided into two parts, for new connections, and for on-going ones. An integrated parameter is defined by the FF algorithm, the final cost factor $F_{u,p,s,r}$, which reflects the degree of adequacy of a given BS/RAT to a given user: it takes each cell of the r RAT for each u user, belonging to the p customer profile, requesting a given s service. The RAT selection algorithm is considered differently, depending on whether the selection is done at session set-up or during an on-going connection.

For a user requesting service s, Fig. 3, the procedure is:

- Measure the $F_{u,p,s,r}$ for each candidate cell k_r of the r detected RAT.
- Select the RAT r having the cell with the highest $F_{u,p,s,r}$ among all candidate cells:

$$R = \arg\max_r \left(\max_{k_r} F_{u,p,s,r}(k_r) \right) \qquad (5)$$

   In case that two or more RATs have the same $F_{u,p,s,r}$ value, then, the less loaded RAT is selected.
- Run the Call Admission Control (CAC) procedure in the RAT r. If admission is not possible, try the next RAT in decreasing order of $F_{u,p,s,r}$, provided that its $F_{u,p,s,r}$ value is higher than 0. If no other RATs with $F_{u,p,s,r}$ higher than 0 exist, block the call.

For on-going connections, Fig. 4, the proposed criterion to execute a VHO algorithm is based on the FF. It assumes that the MT is connected to the RAT denoted as "servingRAT" and cell denoted as "servingCell".

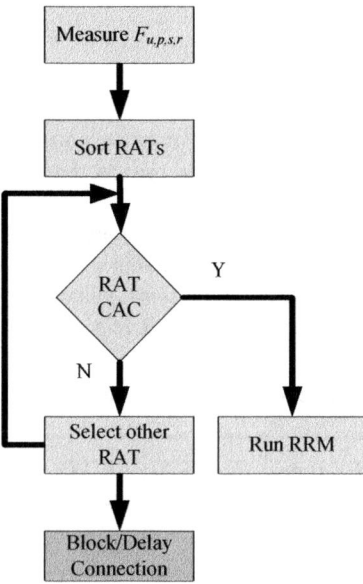

Fig. 3. FF BS/RAT selection algorithm (new connections)

In this case of on-going connections, the procedure is:

- For each candidate cell and RAT, monitor the corresponding $F_{u,p,s,r}$ (k_r). Measures should be averaged during a period T.
- If condition

$$F_{u,p,s,r}\left(k_r\right) > F_{u,p,s,servingRAN}\left(servingCell\right) + \Delta_{VHO} \tag{6}$$

holds during a period T_{VHO}, then a VHO to RAT r and cell k_r should be triggered, provided that there are available resources for the user in this RAT and cell.

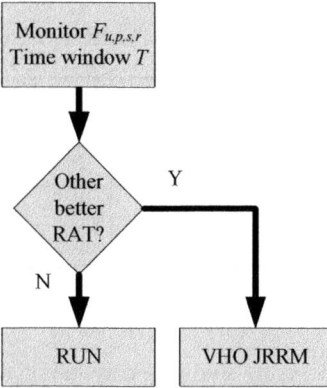

Fig. 4. The FF VHO triggering process (on-going connections)

4   Scenario and Results

4.1   Reference Scenario

In order to assess the impact on EE performance produced by the previous algorithms, micro-cell BS power and traffic models are used to design a reference scenario proposed in 8.

As mentioned previously, the BS equipment power model consumption corresponds to a typical commercially BS in 2010, used in this work as reference for simulations and for system level EE gains. Table 1 presents values for the BS, implementing an approximation to Fig. 1; the BS system power levels are computed via (1).

Table 1. Model Parameters for Micro-Cells

BS type	P_{max} [W]	P_0 [W]	Δ_P
Micro	6.3	106	6.35

Different services are also under test, because different RATs present different radio bearers, since they handle services in a different manner (e.g., voice in circuit switch, or data in packet switch networks). The reference number assumed for the users' density is 1 000 user/km^2, which corresponds to a Low (L) traffic case; in order to assess the traffic load impact on EE, users density is increased by 25 and 50%, corresponding to Medium (M) and High (H) traffic load cases, respectively. In the reference scenario, one assumes two BSs, WCDMA and HSDPA, with a 500 m radius, WCDMA being considered the legacy RAT in which EE may be enhanced. The considered service set is defined according to Table 2, where reference parameters are provided for each service, arrival rate λ, session time τ or volume V_D being defined. The average number of active users defined in the simulator setting is valid for the Low traffic situation and for both BSs.

Table 2. Services and Traffic (Reference)

Services	Average number of active users	λ	τ [s]	V_D [MB]
Voice	15	0.31	90	-
WWW	6	0.5	-	0.1
FTP	6	0.1	-	10

4.2   Results

The first step to obtain EE gains starts by computing the BS RF power level, and comparing two distinguished cases: the previous EE algorithms being OFF and ON.

Based on the reference scenario, different simulations were performed, RF power results being presented in Table 3, where each service is individually evaluated (active users are performing only this particular service), as well for the case where all services are combined.

Table 3. RF Power Levels and Services Traffic Load

		WCDMA RF Power [W] Traffic Load Profiles		
Services	EE [ON/OFF]	Low	Medium	High
Voice	OFF	1.9	2.9	2.9
	ON	1.4	1.5	1.9
WWW	OFF	4.5	4.9	5.3
	ON	2.4	3.0	5.2
FTP	OFF	3.3	3.5	5.4
	ON	2.3	2.7	5.0
ALL	OFF	6.8	6.6	6.2
	ON	3.8	5.3	6.0

The EE gain results for WCDMA G_{RF}, at the radio level, are presented by Fig. 5. Corresponding results, using the previous power model, at BS system level, G_{SL}, are presented in Fig. 6. G_{RF} is calculated by comparing results were the EE is OFF and ON; by using (1), G_{SL} is computed in the same way. By observing Fig. 5, one concludes that only the voice service increases its EE for the Medium load case. This happens because voice has less impact on the network load (compared to data), providing a margin to increase EE even in Medium load; however for High load, voice users trigger VHOs, providing a lower EE gain.

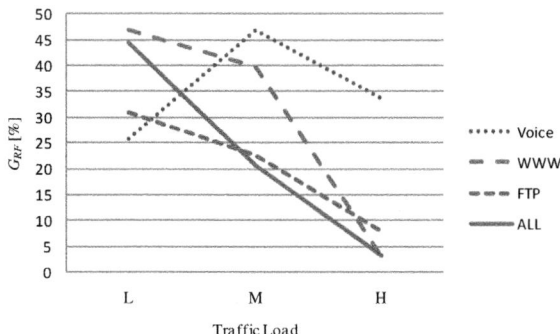

Fig. 5. RF EE gains for each simulated case

When traffic rises, EE gain decreases, because the number of VHOs triggered by the FF also decreases, due to the lower power margin offered by multi-RAT BSs. The only exception is voice, since it takes less radio BSs resources, thus, networks can offer EE gains even when voice traffic increases (e.g., Medium load). Other effects become important, e.g., handover receiving RATs become more loaded, therefore, less power efficient, since more intra-cell interference becomes also significant. Another effect is distance, since HSDPA networks, by comparison with legacy ones, may become less power efficient for long range communications, or may reduce the coverage of high bit rate services, leading potential VHO situations to be less interesting or impossible. For the WWW and FTP services, differences on EE gains curves are due to their session density, duration, or burstiness intrinsic natures. The results from Fig. 6 follow the previous RF gains, however, for the same cases the effect of the ΔP factor can be observed; for example, in the Low case, when all services are combined, it presents higher gains compared to the single WWW case. Anyway, for the all services case, EE ranges from about 1 up to 13%, which means that one can cut 13% in WCDMA CO_2 BSs emissions and network operation costs.

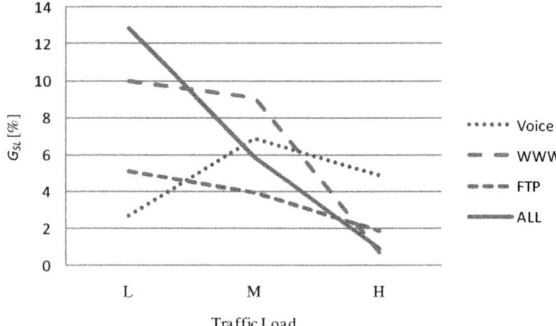

Fig. 6. EE at system level gains for each simulated case

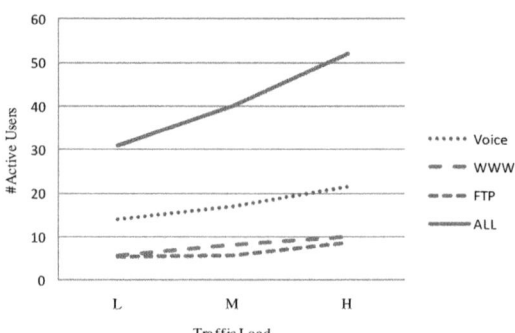

Fig. 7. Average number of active MTs/service generated

When all services are simulated simultaneously, the RF EE gain decreases with the traffic load, but it can achieve relatively high gains in the Low traffic case, because, in this traffic scenario, VHOs have some power margin, producing EE gains (being coherent with WWW and FTP trends).

Fig. 7 shows the total average number of active users generated by the tool for both BSs under simulation and for all previous situations. The FTP and WWW results are very similar as expected. These results confirm the previous effects, since it is shown that real-time services rise according to the plan defined in the scenario.

5 Conclusions

This paper proposes VHOs to be used as one more technique in the battle for power consumption reduction in mobile cellular infrastructures, being applicable in a multi-RAT environment. Simulations include propagation issues, services and traffic models and JRRM algorithms to trigger VHOs guided by EE metrics, such as W/bps. The capability to distinguish packets overhead and users' data is also included.

The model gives EE gains directly at the RF level, using the presented BS power model, being possible to compute EE gains at the BS system level.

Results for voice, web browsing, and file transfer show EE gains at RF levels ranging from 5 up to 46%, for High and Low traffic cases, respectively.

EE gains for micro-cell BS power consumption, at the system level, range from 1 to 13%, achieved for High and Low traffic cases, respectively. Thus, one may conclude that in very high traffic load conditions, VHOs will have a very short margin to decrease power consumptions, since BSs will be closer to congestion. Nevertheless, for Low and Medium traffic loads, EE gains present interesting values that should be considered.

Acknowledgements. The research leading to these results has received funding from the European Community's Seventh Framework Programme FP7/2007-2013 under grant agreement n° 247733 – project EARTH.

References

1. http://www.ict-EARTH.eu/ (February 2011)
2. Ericsson, Sustainable energy use in mobile communications, White paper (August 2007), http://www.ericsson.com/campaign/sustainable_mobile_communic ations/downloads/sustainable_energy.pdf
3. Desset, C., Ahmed, N., Dejonghe, A.: Energy savings for wireless terminals through smart vertical handover. In: Proc. of ICC 2009 - 2009 IEEE International Conference on Communications, Dresden, Germany (June 2009), http://www.ieee-icc.org/2009
4. Radwan, A., Rodriguez, J., Gomes, A., Sá, E.: C2POWER Approach for Power Saving in Multi-standard Wireless Devices. In: Proc. of VTC 2011 Spring – 73rd IEEE Vehicular Technology Conference, Budapest, Hungary (May 2011), http://www.ict-c2power.eu

5. Kanno, I., Yamazaki, K., Ikeda, Y., Ishikawa, H.: Adaptive Energy Centric Radio Access Selection for Vertical Handover in Heterogeneous Networks. In: Proc. of WCNC 2010 – IEEE Wireless Communications and Networking Conference, Sydney, Australia (April 2010), http://www.ieee-wcnc.org/2010
6. Seo, S., Song, J.S.: Energy-Efficient Vertical Handover Mechanism. IEICE Transactions on Communications E92-B(9), 2964–2966 (2009)
7. Serrador, A. (ed.): Simulation tools: final version capabilities and features, IST-AROMA Project, Del. D13, EC-IST Office, Brussels, Belgium (April 2007), http://www.aroma-ist.upc.edu
8. Imran, M.A., Katranaras, E. (eds.): Energy efficiency analysis of the reference systems, areas of improvements and target breakdown. ICT-EARTH Project, Deliverable D2.3, EC-IST Office, Brussels, Belgium (January 2011), http://www.ict-EARTH.eu
9. Serrador, A., Correia, L.M.: A Cost Function Model for CRRM over Heterogeneous. Wireless Personal Communications 59(2), 313–329 (2011), http://www.springer.com/engineering/signals/journal/11277
10. Serrador, A., Correia, L.M.: Policies For a Cost Function For Heterogeneous Networks Performance Evaluation. In: Proc. of PIMRC 2007 – 18th Annual IEEE International Symposium on Personal Indoor and Mobile Radio Communications, Athens, Greece (September 2007), http://www.pimrc2007.org
11. Pérez-Romero, J.: Operator's RAT Selection Policies Based on the Fittingness Factor Concept. In: Proc. of 16th IST Mobile and Wireless Communications Summit, Budapest, Hungary (June 2007)

Inter-Domain Route Diversity for the Internet

Xavier Misseri[1], Ivan Gojmerac[2], and Jean-Louis Rougier[1]

[1] TÉLÉCOM ParisTech, 46 Rue Barrault, 75013 Paris, France
[2] FTW – Telecommunications Research Center Vienna, Donau-City-Str. 1,
1220 Vienna, Austria
{misseri,rougier}@telecom-paristech.fr, gojmerac@ftw.at

Abstract. The current inter-domain routing in the Internet, which is based on BGP-4, does not allow for the use of multiple paths, but rather restricts the routing to a single path for each destination prefix. This fact is especially unfortunate considering the vast route diversity which is inherently present in the global Internet graph. Therefore, we propose Inter-Domain Route Diversity (IDRD) as an overlay mechanism which enables efficient, backwards compatible and incrementally deployable introduction of route diversity in the Internet. Beyond presenting the architecture of IDRD, this paper also presents the conditions which ensure the stability of the proposed mechanism as a fundamental prerequisite for its deployment in real-world scenarios.

Keywords: Inter-domain routing, BGP, Path diversity, IDRD, Map-and-Encap.

1 Introduction and Related Work

In order for the Internet to function as a set of individual networks belonging to different administrative domains, a mechanism is required which will provide for the global exchange of routing information. This role is currently fulfilled by the Border Gateway Protocol (BGP) [13], which propagates IP prefix reachability information by exchanging so-called *path vectors* between neighbor networks. In order to achieve scalability in such an approach, each domain selects only a single route (i.e., *path*) per IP prefix, and accordingly, only the selected path is propagated to the neighbors. Thereby, the inherent Internet-scale route diversity is not put into use, as effectively only single-path routes towards any global subnetwork are enabled. This impedes multi-path gains even at the level of Tier-1 networks, in spite of the vast diversity they are exposed to (cf. [16]).

Additionally, BGP route selection represents a stumbling block for the efficient and flexible operation of individual domains, as the *BGP decision process* (which determines the next hop neighbor network for each global prefix) is comprised of a number of successive, hard-coded and static rules based on comparisons of global or local path attributes, such as the *LOCAL_PREF*, *AS Path* length or the *MED* (Multi-Exit Discriminator), eventually always ending in a tie-break which determines the single path used.

Z. Becvar et al. (Eds.): NETWORKING 2012 Workshops, LNCS 7291, pp. 63–71, 2012.

Nevertheless, BGP does encompass some potential for traffic engineering, with several techniques having been proposed in literature. For instance, [12] proposes Local-Preference tweaking mechanisms in order for multihomed ASes to control their outbound traffic sent towards the different providers. However, BGP traffic engineering techniques are quite limited and coarse grained, as extensively laid out in [3].

On a similar note, some work has been performed in the field of inter-domain path diversity, however, a simple and effective solution for putting the multiple viable paths into use has not yet been formulated.

In [15], BGP add-path is presented as an extension of BGP that allows for the announcement of several paths per prefix, but which at the same time does not change the BGP decision process, effectively resulting in a good technique for fast failover in combination with BGP Route Reflectors (RRs).

In order to enable simultaneous use of multiple routes, the *Multipath BGP* extension [1,8] slightly changes the BGP decision process. While *Multipath BGP* allows for balancing traffic in equal shares among the available paths, it does not propagate path diversity to its neighbors, and furthermore it is subject to tough constraints, as the different routes must be very similar concerning their metrics (i.e., Local-Preference, AS path length and MED).

[18] proposes Multi-path Interdomain ROuting (MIRO) as an architecture for the selection, export and enforcement of alternative inter-domain routes. Whereas MIRO does address some problems discussed in the present paper, at the same time it does not provide a stability analysis for the proposed scheme, which we consider to be a prerequisite for any practical deployment. And concerning message exchange, MIRO specifies that routes are *pulled* and not *pushed*; while we do understand this paradigm in the context of scalability issues, we believe that the solution should rather lie in the selection of routes which are to be pushed in the first place, than in abandoning the push concept altogether.

Last, but not least, [17] proposes D-BGP and B-BGP as two interdomain routing proposals that propagate path diversity. The goal of these proposals is to speed-up the recovery of BGP by propagating a maximally disjoint alternative path associated to the BGP best route. The propagated multi-paths are however only used for back-up purposes, and the proposed changes are supposed to be integrated into BGP, which we do not find to be realistic.

In the next section, we will advance the current state of the art by introducing Inter-Domain Route Diversity (IDRD) which allows ISPs to propagate more routes towards their customers, providers and peers. Moreover, we propose to relax the constraints on the BGP selection process, thereby mainly focusing on the *prefer client routes* conditions. Overall, the propagation of the multiple routes faces several challenges, foremostly in assuring the stability of the control plane. Indeed, the use and the propagation of diversity has substantial significance only if the domains can select policies that are different from the BGP decision process. E.g., a domain should be able to propagate routes that do not have the highest *LOCAL_PREF*. However, we recognize that the Valley Free conditions from [4], which ensure stability in the current Internet, are not sufficient in this case, but that additional mechanisms are required for ensuring the stability of the propagated diversity.

The rest of this paper is structured as follows: Section 2 presents the architecture of IDRD as our proposed solution, thereby equally focusing on control plane and data plane aspects, as well as on the prospective use cases, architecture deployability and its backward compatibility. Subsequently, Section 3 provides an analysis of IDRD stability before Section 4 concludes the paper with a summary of the main results.

2 Architecture Proposal for Route Diversity

In this section, we present Inter-Domain Route Diversity (IDRD), which enables the use of the inherent topological diversity present in today's Internet. More specifically, we describe the IDRD control and data planes, followed by a discussion of the most relevant IDRD use cases. Finally, we provide strong arguments for the real-world deployability of our solution due to several key properties of IDRD.

Before presenting IDRD in the next subsections, here we aim at providing clarity on the character of this solution, i.e., we wish to stress that IDRD by no means aims at replacing BGP. Instead, we see IDRD as an add-on to the present Internet, which can be deployed by some domains in parallel to BGP as a higher-layer overlay.

In order to make the propagation (control plane) and the use (data plane) of diversity possible, we base our architecture on the *map-and-encap* paradigm, which allows the traffic to be encapsulated according to the parameters provided by a Mapping System (cf. [10] for an existing example). The Mapping System (MS) can either be internal or external with respect to the encapsulating routers, and structurally it can either follow a centralized or a decentralized logic.

2.1 IDRD Control Plane

With IDRD, each domain stores the information on path diversity within its own Mapping System (MS). As a domain that has adopted IDRD may be connected to domains that have not adopted this architecture, the routing information coming from these neighboring domains (in the form of eBGP updates) can be redistributed into the MS. Conventional BGP is thus a source of diversity in such a case.

The propagation of multiple paths is performed at the MS-level for neighbors that have adopted the architecture, i.e., their MSes communicate directly in order to provide route diversity. This diversity information contains BGP metrics and may also contain other metrics, e.g., price, capacity, etc. Once the set of multiple paths has been received (either via BGP redistribution or via inter-MS communication), the domain can select a subset of those paths which it finds interesting. The MS can compute advanced selection policies based on price, stability, political relationships, etc.

Once a set of paths is selected, it is propagated by the domain to its neighbors. In this context we note that the paths selected by a domain are not necessarily all put into use. Instead, the selected set of paths represents an assurance that neighbor domains can utilize them using the map-and-encap scheme.

The proposed approach deeply relaxes the 'prefer client' constraint of BGP due to the multiple potential paths which can be conveyed and used. However, it is important to note that all traffic which is received outside of the map-and-encap scheme (i.e., all *conventional* traffic) will still be forwarded via the standard BGP best routes for reasons of backward compatibility.

Once the diversity is propagated between ASes, the Autonomous System Border Routers (ASBRs) must be made aware of the corresponding mapping information. The MS can either directly push the mapping or await mapping requests from neighbor ASBRs. Each MS entry provided to an ASBR contains at least the following information:

- The association between the flow identifier and the *next hop* / ASBR,
- The association between incoming and outgoing flow identifiers (cf. Sec. 2.2).

2.2   IDRD Data Plane

Life-cycle of a Packet: In order to implement the usage of alternative paths advertised by IDRD in the present Internet, we propose to apply packet encapsulation, similarly to the scheme presented in [7]. There are two different areas of path enforcement which must be taken into account:

- **Intra**-domain path enforcement: When a packet arrives at a domain entrance, the ASBR asks the mapping system about which exit ASBR the packet must be forwarded to. According to its *flow identifier*, the Mapping System (MS) specifies the exit ASBR. The entry ASBR then encapsulates the packet and forwards it to the correct exit ASBR. It is important to note that the encapsulation scheme is local and that it has got no impact on neighboring domains. Therefore, each AS can individually choose its encapsulation scheme (e.g., IPv4, MPLS, etc.). Finally, once the packet arrives at the exit ASBR, it gets decapsulated.
- **Inter**-domain path enforcement: When arriving at the exit ASBR, a packet gets encapsulated in order to enforce its path towards the next domain's ASBR. As in the case of intra-domain path enforcement, the mapping resolution can be either pushed or pulled. But in contrast to the intra-domain case, the inter-domain encapsulation scheme must be negotiated between neighboring ASes in order to be inter-operable.

Flow Identification: The ASBRs must forward packets from one tunnel to next. As several paths are available in order to reach a destination IP prefix, the destination IP address in the inner IP header is no longer sufficient for making the forwarding decision. Therefore, in addition to the inner IP destination address (the real destination host), an identifier can be used to specify the route which is to be used. In order to be scalable, this identifier must be assigned and used locally in each domain or at each peering/transit link. Inter-AS and intra-AS identifiers of the same path must be aligned in order to be able to choose a coherent path. ASBRs must then be able to swap incoming identifiers with outgoing identifiers.

2.3 IDRD Use Cases

The use cases for inter-domain route/path diversity are well-known and we only briefly enumerate them here. Firstly, route diversity has the potential to increase the overall network capacity between two points in the Internet, i.e., it can be used for traffic engineering and load balancing (cf. [3,9]). Secondly, having available a set of multiple disjoint paths can also be used for increasing resilience (cf. [19]). Both mentioned benefits could potentially also apply to end customers who employ layer-4 path diversity schemes, like e.g. MP-TCP [6]. And finally, flexible and explicit route enforcement represents an important tool for inter-domain Quality of Service (QoS) mechanisms, which will substantially gain importance if large-scale capacity overprovisioning in the Internet becomes unfeasible (cf. [2]).

2.4 Backward Compatibility and Deployability of IDRD

The design of IDRD repsects both successful protocol design requirements postulated by C. Dovrolis in [14]. Firstly, our architecture is *backward compatible*, as it is incrementaly deployable among only a subset of Internet Service Providers (ISPs) due to its seamless compatibility with the current Internet. Secondly, IDRD is *incrementally deployable* in the sense that it brings benefits to its early adopters even if not broadly deployed. Therefore, we believe that IDRD has the potential to achieve practical relevance in the mid-term future.

3 Discussion of IDRD Stability

3.1 Instability Example

Enabling the propagation of path diversity in the current Internet may lead to oscillations. Figures 1 and 2 provide a simple but stunning example for the instabilities which might occur: AS_A, AS_B and AS_C are peers, and AS_D is the client of the other three ASes. In order to be able to reach AS_D via multiple paths, each AS selects a second route according to local policies, in addition to the first (i.e., direct) BGP best route. Each AS uses a local decision process and ranks the potential paths for the alternative route choice. In our example, AS_A orders the alternatives by priority as ACD, AD, ABD. Figure 2 presents the stepwise change of path selection in each AS. Thereby, the selections which have changed since the last step are highlited in red, the selections which are being propagated to neighbors are underlined, and paths which are being unselected (withdrawn) are crossed out. According to the previously listed priority list, if AS_A receives the path CD from C (as in Line 3 of Figure 2), it chooses the path ACD and withdraws the path AD. And concurrently, AS_A sends the withdrawal

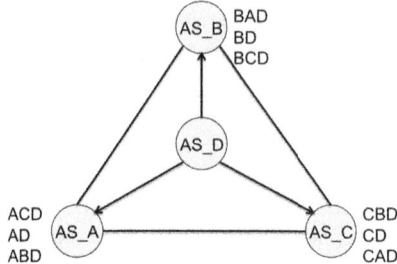

Step	AS_A	AS_B	AS_C
1	AD	BD	C̲D̲
2	ACD(A̶D̶)	BD	CD
3	ACD	B̲D̲	CD
4	ACD	BD	CBD(C̶D̶)
5	A̲D̲	BD	CBD
6	AD	BAD(B̶D̶)	CBD
7	AD	BAD	C̲D̲
8	ACD(A̶D̶)	BAD	CD
9	ACD	B̲D̲	CD

Oscillation !!!

Fig. 1. Example topology   **Fig. 2.** Evolution of the routing decisions

of AD to its neighbors, however, it does not propagate the newly selected path ACD to its peers (due to adherence to the Valley Free conditions [4], as discussed in Section 3.2).

We can see from Figure 2 that Steps 3 and 9 are identical, which implies that the system will enter into oscillations (cf. also [5] for further examples of instabilities in inter-domain routing). The next subsection underlines the sufficient conditions to reliably avoid oscillations and it provides a pointer to a mathematical proof of IDRD stability.

3.2 IDRD Stability Conditions

In the above example we have highlighted that the propagation of multiple routes can lead to oscillations. Therefore, in this part we present a criterion which – in addition to the the Valley Free conditions [4] – ensures the stability of IDRD. In the IDRD architecture, a domain d receives:

- A set of routes coming from clients (i.e., client diversity): $R_{d,c}$ (where d denotes a *domain* and c denotes the whole set of *clients*).
- A set of routes coming from peers and providers (i.e., peer/provider diversity): $R_{d,p}$ (where p denotes the whole set of *peers and providers*).

The domain d uses a decision process λ_d to select interesting routes among the two sets of route candidates: $S_d = S_{d,c} \cup S_{d,p} = \lambda_u(R_{d,c} \cup R_{d,p})$ where S_d denotes the set of routes selected by the domain d, $S_{d,c}$ stands for the set of selected client routes, and where $S_{d,p}$ denotes the set of selected peer/provider routes.

In the current BGP-based Internet, Valley Free policies already apply and they ensure its stability. In terms of IDRD, the well known Valley Free conditions translate to:

- Each d sends only the set of selected client routes ($S_{d,c}$) to its neighbor peers and/or providers.
- Each d sends all selected routes ($S_{d,c} \cup S_{d,p}$) to client neighbors.

In order to ensure the global stability of IDRD, the following stability criterion introduces a strong requirement in addition to the two Valley Free conditions stated above.

> IDRD Stability Criterion: Routes received from peers and providers must have no impact on the selection of routes received from clients. More formally, if we have $S_d = S_{d,c} \cup S_{d,p} = \lambda_d(R_{d,c} \cup R_{d,p})$, then $S_{d,c}$ must be independent from $R_{d,p}$.

We can analyse the impact of this criterion on the oscillation example given in Section 3.1. In Steps 2, 4, 6 and 8, the ASes have received a route from a peer, subsequently unselecting the route coming from their client. However, this is strongly prohibited by the stated stability criterion, which ensures that the ASes select and propagate client routes independently of peer and provider routes. In our example, ASes must thereore select both peer and client routes. Nevertheless, for the traffic originated from within the domain, each AS can still opt for using only one of the advertised routes.

Due to limited space in this paper, for a comprehensive proof of IDRD stability we refer the reader to our technical report in [11]. There we prove that an IDRD system that respects the previously stated criterion and the Valley Free conditions is safe, meaning that it converges to a stable state from any initial state and that this stable state is unique. We prove this statement in a three step procedure. The first step proves that an IDRD system that respects the previous criterion and the Valley Free conditions has a stable state, and that this stable state is unique. The second step proves that it reaches the stable state for any initial state. Finally, the third step proves that this stable state is reached within a finite time interval.

4   Conclusions and Future Work

Due to its distributed nature, the global Internet displays vast potential path diversity. However, for stability reasons, BGP-4 as the current inter-domain routing protocol in the Internet does not enable its utilization, as it allows only for a single path towards each destination prefix.

In order to mitigate this restriction, in this paper we propose Inter-Domain Route Diversity (IDRD) which allows for the propagation (control plane) and the use (data plane) of the present Internet path diversity. In order to achieve optimal backward compatibility as well as incremental deployability, we have designed IDRD as an overlay mechanism which can operate in parallel to BGP, and which enables partial deployment at the AS-level. As far as use cases for IDRD are concerned, we identify substantial potential in the areas of traffic engineering and load balancing, which aim at optimal utilization of the available network capacities. Furthermore, we believe that explicit selection of multiple end-to-end paths using IDRD can play an important role in the provisioning of QoS-enabled Internet services, as well as in the improvement of Internet-wide resilience in the presence of anomalies and component failures. After presenting IDRD in detail, we have paid great attention to the issue of *stability*, which we consider to be the most important criterion when introducing any new protocol to the Internet. Accordingly, in Section 3 we provide an in-depth discussion of this topic, accompanied with pointers to our comprehensive proof of IDRD stability.

Concerning future research on IDRD, our work is far from coming to an end. Next, we aim at further detailing the IDRD architecture and devising advanced decision processes for route selection. Furthermore, we also intend to provide a quantitative analysis of the amount of path diversity in the present Internet.

Acknowledgments. This work has received funding from the European Unions's Seventh Framework Program (FP7/2007-2013) under grant agreement 248567 for the ETICS project (https://www.ict-etics.eu/). FTW is funded within the COMET Program by the Austrian Government and the City of Vienna.

References

1. Cisco Systems: BGP Best Path Selection Algorithm, http://www.cisco.com/en/US/tech/tk365/technologies_tech_note09186a0080094431.shtml
2. ETICS Project: Website (2012), http://www.ict-etics.eu
3. Feamster, N., Borkenhagen, J., Rexford, J.: Guidelines for Interdomain Traffic Engineering. ACM SIGCOMM Computer Communication Review 33(5) (2003)
4. Gao, L., Rexford, J.: Stable Internet Routing Without Global Coordination. IEEE/ACM Transactions on Networking 9(6), 307–317 (2000)
5. Griffin, T.G., Shepherd, F.B., Wilfong, G.: The stable paths problem and interdomain routing. IEEE/ACM Transactions on Networking (TON) 10(2), 232–243 (2002)
6. Handley, M., Raiciu, C., Ford, A., Barre, S., Iyengar, J.: Architectural Guidelines for Multipath TCP Development. IETF RFC 6182 (Draft Standard) (2011)
7. Jiayue, H., Rexford, J.: Toward Internet-Wide Multipath Routing. IEEE Network 22(2), 16–21 (2008)
8. Juniper: Configure BGP to Select Multiple BGP Paths (2012), http://www.juniper.net/techpubs/software/junos/junos53/swconfig53-ipv6/html/ipv6-bgp-config29.html
9. Kostopoulos, A., Warma, H., Leva, T., Heinrich, B., Ford, A., Eggert, L.: Towards Multipath TCP Adoption: Challenges and Opportunities. In: 6th EURO-NF Conference on Next Generation Internet (NGI), pp. 1–8 (June 2010)
10. Lewis, D., Fuller, V., Farinacci, D., Meyer, D.: Locator/ID Separation Protocol (LISP), IETF Internet Draft (Work in Progress) (January 2012)
11. Misseri, X., Gojmerac, I., Rougier, J.L.: Technical report – Stability of Global Diversity Propagation (2012), http://perso.telecom-paristech.fr/misseri/Files/10-Stability-Diversity-Propagation-Technical-report.pdf
12. Quoitin, B., Pelsser, C., Swinnen, L., Bonaventure, O., Uhlig, S.: Interdomain Traffic Engineering with BGP. IEEE Communications Magazine 41(5), 122–128 (2003)
13. Rekhter, Y., Li, T., Hares, S.: A Border Gateway Protocol 4 (BGP-4). IETF RFC 4271 (Draft Standard) (2006)
14. Rexford, J., Dovrolis, C.: Future Internet Architecture: Clean-Slate Versus Evolutionary Research. Communications of the ACM 53(9), 36–40 (2010)
15. Scudder, J., Retana, A., Walton, D., Chen, E.: Advertisement of Multiple Paths in BGP, IETF Internet Draft (Work in Progress) (September 2011)

16. Uhlig, S., Tandel, S.: Quantifying the BGP Routes Diversity Inside a Tier-1 Network. In: Boavida, F., Plagemann, T., Stiller, B., Westphal, C., Monteiro, E. (eds.) NETWORKING 2006. LNCS, vol. 3976, pp. 1002–1013. Springer, Heidelberg (2006)
17. Wang, F., Gao, L.: Path Diversity Aware Interdomain Routing. In: IEEE INFOCOM 2009, pp. 307–315 (2009)
18. Xu, W., Rexford, J.: MIRO: Multi-path Interdomain ROuting. In: Proc. ACM SIGCOMM 2006, Pisa, Italy, pp. 171–182 (September 2006)
19. Yannuzzi, M., Masip-Bruin, X., Sanchez, S., Domingo-Pascual, J., Orda, A., Sprintson, A.: On the Challenges of Establishing Disjoint QoS IP/MPLS Paths Across Multiple Domains. IEEE Communications Magazine 44(12), 60–66 (2006)

A Proposal of Business Model Design Parameters for Future Internet Carriers

Antonio Ghezzi

Politecnico di Milano,
Department of Management,
Economics and Industrial Engineering
via Lambruschini 4B, 20156 Milan, Italy
Antonio1.ghezzi@polimi.it

Abstract. Future Internet evolution requires innovative strategic stances and the design of original business models from actors involved in the ecosystem. The study focuses on Internet Carriers, recently striving to make their business sustainable, and proposes to enclose in a single reference framework all the critical levers, either consolidated or innovative, such actors can employ in order to design their value proposition, value network integration, and financial configuration. The framework grounds its findings on multiple case studies, and, by presenting an insightful list of business model parameters for Carriers, sheds light on key emerging strategic and tactical trends in the Internet interconnections market.

Keywords: Future Internet, Carriers, Business Model Design, Strategy, Interconnections.

1   Introduction

The Internet's future development is not only heavily depending on its technological evolution, but also on business sustainability for the interconnection ecosystem the Web relies on, where various players characterized by fairly different economics are coexisting [3].

Current Internet technologies and business rules for network interconnection are proving to be no longer able to support the sustainable development of all actors along the value network, i.e. from the application to the network service providers. While Over-The-Top (OTT) providers develop high performance applications that create new revenues opportunities for them, Carriers have to cope with more and more constrainable traffic they cannot control and charge in order to recover corresponding extra network investment costs [18]. Indeed, peering agreements remain static and insensitive with respect to Quality of Service parameters, and operators rely mainly on the revenues coming from flat rate pricing in access networks.

To lighten this quite gloomy picture, Internet Service Providers (ISP) or Carriers are required to elaborate innovative strategies, in turn resulting in original tactical choices at a business model design level. The present study, through collecting the multifaceted perspectives of 10 Carriers (integrated by the complementary view of 5

Z. Becvar et al. (Eds.): NETWORKING 2012 Workshops, LNCS 7291, pp. 72–79, 2012.

OTTs), aims at disclosing such strategic and tactical decisions (either already consolidated in the current market practices, or innovatively derived from original stances) and present them in a comprehensive framework for to support Carriers in the business model design process from a Future Internet perspective.

2   Methodology

The research leverages on two main methodological pillars: literature review and multiple case studies. The literature review focused on the significant and quite innovative stream of the broad Strategic Management theory. i.e. Business Model Design, posing the theoretical bases for the proposed Carriers model's building blocks.

With reference to the empirical research methodology, case studies are defined [15] as "empirical inquiries that investigates a contemporary phenomenon within its real-life context, especially when the boundaries between phenomenon and context are not clearly evident; and in which multiple sources of evidence are used". Qualitative research methodology is particularly suitable for reaching the research objectives, which aim at understanding the complex phenomenon of Business Model design within a given industry – that is, the Internet Interconnections Industry – characterised by a high level of dynamicity and competitive turbulence, and at thus building new theory, or extending existing theories, on such context [13].

To accomplish the previously identified research propositions, exploratory case studies based on semi-structured interviews are performed on a set of 15 companies operating in the broad Internet Interconnection market, and belonging to two major categories: Carriers (10 cases: British Telecom; Deutsche Telekom; France Telecom; Telecom Italia; Telefonica; Telenor; Fastweb; Tiscali; Colt; Libero Infostrada) and OTT/Content Providers (5 cases: Akamai; Google; Yahoo, Buongiorno, Dada).

The interviews focused on addressing which are the most critical choices to be made at a Business Model design level for Carriers. The semi-structured nature of the interview allows to start from some key issues identified through the literature, but also to let any innovative issue emerge from the open discussion.

As argued in [7], a multiple case studies approach reinforces the generalisation of results obtained, and allows to perform a cross analysis on Business Models characteristics and their combinations, due to the presence of extreme cases, polar types or niche situations within the theoretical sample. As the validity and reliability of case studies rest heavily on the correctness of the information provided by the interviewees and can be assured by using multiple sources or looking at data in multiple ways, several sources of evidences or research methods are to be employed: interviews (to be considered the primary data source), analysis of internal documents (both official and informal), study of secondary sources on the firm (i.e. research reports, websites, newsletters, white papers, databases, and conferences proceedings). This combination of sources allows obtaining perceptual triangulation, essential for assuring rigorous results in qualitative research.

3   Literature Review

The concept of business model generally refers to the "architecture of a business" or the way firms structure their activities in order to create and capture value [10] [12] [14].

As a literature stream, business model design has evolved from a piecemeal approach that looked for the single identification of typologies or taxonomies of models [1] [10] [11] [14] to one searching for the development of a clear and unambiguous ontology [9] – that is, the definition of the basic concepts of a theory – that could be employed as a generalized tool for supporting strategy analysis of firms. In parallel, business model has become an extensive and dynamic concept, as its focus has shifted from the single firm to the network of firms, and from the sole firm's positioning within the network to its entire interrelations and hierarchies [2].

It is widely accepted in literature that a business model shall be analyzed through a multi-category approach, as a combination of multiple design dimension, elements or building blocks. However, the proposed dimensions and interrelations are quite diverse, and the existing body of knowledge shows a lack of homogeneity [6].

Noteworthy attempts of providing a unified and consistent framework can be found in several works. Yu [16] mentions different critical business model components such as assets, markets, customers, competitors, products, services, costs, prices, revenues, profits, market shares, economic scales, marketing strategies, competitive advantages. According to Hedman and Kalling [5], the conceptual business model should include elements such as customers and competitors, the offering, activities and organization, resources and factor market interactions. Osterwalder, in his proposed business model design template [9], identifies four key dimensions for a business model: infrastructure; offering; customers; and finance. Morris and others [8] propose a six-component framework for characterizing a business model, regardless of venture type, which comprehends: value creation; value target; internal source of advantage; firm market positioning; value capture; entrepreneur's time, scope and size ambitions. Ballon [2] holds that the recurrent parameters a business model is built on can be brought back to the general concepts of value – value proposition and financial configuration – and control – inter-firm or value network relationships –. For Johnson and others [6] a business model consists of four interlocking elements that, taken together, create and deliver value: customer value; profit formula; key resources; and key processes. Recently, Zott and Amit [17] suggest two sets of parameters that systems designers need to consider: design elements – content, structure and governance – that describe the architecture of an activity system; and design themes – novelty, lock-in, complementarities and efficiency – that describe the sources of the activity system's value creation.

Notwithstanding the presence of several alternative frameworks for business model design, the literature review allows to claim most studies converge and focus on the concept of value: either it is the value a firm plans to offer to its customers (i.e. value proposition); or the value it creates, shares and competes for with other external actors in its market, depending on the level and nature of its value network integration; or, in the end, the value as a result of the proper selection of a cost structure and a revenue model [4].

4   Business Model Design Parameters for Future Internet Carriers

The research carried out through the multiple case studies allowed to shed light on the core business model design parameters for Carriers in the Internet Interconnections market. The findings are synthesized in the BM Framework below provided, which identifies three macro-dimensions. The macro-dimensions are in turn divided into 12 parameters. The framework is mainly directed to Carriers or Internet Service Providers (ISPs) as the referencing actor.

1. ***Value Proposition Parameters.*** It addresses the firm's strategic positioning by considering key business dimensions.
   (a) Product/Service delivered (basic connectivity vs. ASQ vs. Content)
   (b) Target customer (Content Provider/OTT vs. End User)
   (c) Customer value (basic connectivity vs. Assured Service Quality vs. Content)
   (d) Resources and Competencies (technology-oriented vs. content-oriented)
2. ***Value Network Parameters.*** It addresses the firm's level of integration within its referencing Value Network system.
   (a) Vertical Integration (Infrastructure layer vs. Internet Service Layer)
   (b) Customer Ownership (Direct vs. Indirect)
   (c) Interconnection modality (prevalent: transit vs. peering)
   (d) Content delivery model (Client-Server vs. Cloud vs. CDN)
3. ***Financial Configuration Parameters.*** It addresses the firm's configuration generating revenue and cost streams.
   (a) Revenue Model (Single Transaction vs. Subscription)
   (b) Revenue Sharing Model (Present vs. Absent)
   (c) Traffic Charging scheme (Receiving Party Network Pays model vs. Congestion-charging model vs. Sending Party Network Pays model)
   (d) Cost Model (concentrated vs. distributed)

For each and every parameter, the "value range" is identified, i.e. the extremes values or key alternatives the variables can assume, which also represent the major trade-off between opposite choices; the main strategic implications deriving from alternative parameters adoption are also discussed.

Table 1. Business Model Design Parameters for Future Internet Carriers

	Business Model Parameter	Value Range (Trade-off)	Strategic Implications
Value Proposition	*Product/Service Delivered*	Basic connectivity	Traditional business for ISPs. Easier diffusion/substitution.
		Assured Service Quality (ASQ)	Higher potential margins from connectivity. Service differentiation. Two-tiered internet. Net Neutrality broken.
		Content	Traditional business for CP/OTT. Higher margins from content market making. Higher complexity, ISP business diversification
	Target Customer	Content Provider/OTT	Comparable relative bargaining power. High data traffic to/from single customer Peering agreements potentially required.
		End user	Higher relative bargaining power. Low data traffic to/from single customer. Investment in Access Network (last mile).

Table 1. (*continued*)

	Customer Value	Basic connectivity	Lower expenses for meeting customer requirements. Lower service differentiation potential
		Assured Service Quality (ASQ)	Higher expenses for meeting customer requirements (ASQ pipe) Higher service differentiation potential
		Content	Coverage of Content Management activities. Higher margins from content market making. Higher complexity, ISP business diversification
	Resources & Competencies	Technology-oriented	Disposition towards technology partnership.
		Content-oriented	Disposition towards editorial partnership for service creation.
Value Network	Vertical Integration	Infrastructure Layer coverage	Relegation to technology enabler role. Focus on infrastructural investments, network operation and capacity management.
		Internet Service Layer coverage	More invasive role within the Value Network. Investments in both network infrastructure and content management. Potential competition ISP-CP/OTT.
	Customer Ownership	Intermediated	Increased dependence on CP/OTT. Indirect revenue flows.
		Direct	More central role in the Value Network, direct revenues. Potential Competition with CP/OTT
	Interconnection Modality	Transit prevalence	Indirect interconnection. Lower transaction costs for agreement setting. Higher risk of opportunistic behaviour in traffic management. Need for compensation.
		Peering prevalence	Direct interconnection of peers. Higher transaction costs for peering agreement. Higher interconnection efficiency. Lower need for compensation.
	Content-Data delivery model	Client-server	Basic data delivery model. Simpler model. No distribution of intelligence.
		Cloud	Pool of virtualized resources. Lower cost of resource management. Higher scalability and flexibility. Introduction of the Cloud Provider in the Value Network.
		Content Delivery Network (CDN)	Content distribution/storage/management. Improved reliability, throughput, origin server load balancing; lower latencies for consumers. Introduction of the CDN Provider in the Value Network.

Table 1. (*continued*)

Financial Configuration	*Revenue Model*	Single Transaction	One-shot revenues for connectivity consumption and/or content purchasing. Higher margins for single transaction. No customer lock-in.
		Subscription	Flat rate with/without time/traffic/usage/n° downloads caps. Lower margins for single transaction. Customer lock-in and future revenues assured.
	Revenue Sharing Model	Present	Business sharing (opportunities/risks) between ISP-CP/OTT-End user
		Absent	Clear separation between ISP and ISP-CP/OTT-End user businesses
	Traffic Charging scheme	Receiving Party Pays	Traditional charging scheme favouring CP/OTT Lower incentives to invest for ISPs.
		Sender Party Pays	Incentives alignment: ISPs to invest in new capacity and QoS; CP/OTT to use network capacity efficiently and keep traffic on net.
		Congestion Charging	Charging based on network congestion caused. Coverage of ISPs' sunk investment + marginal cost of capacity.
	Cost Model	Concentrated Investment	Increased independence. Increased risk.
		Joint Investment	Risk sharing Increased dependence on partnering actors.

5   Discussion and Conclusions

The framework proposed for supporting the business model crafting for Future Internet Carriers discloses a number of emerging strategic and tactical trends ongoing within the Internet Interconnection market.

Carriers' value proposition is moving from basic connectivity provisioning to the elaboration of an ASQ offer towards end users or OTTs/CPs, in order to incorporate a larger share of the overall market value. To avoid being gradually left aside in a more and more peripheral value network position, Carriers should also strive to maintain direct customer ownership and a coverage of the Internet value added services layer, by becoming enablers and partners of the OTTs/CPs services offer which requires higher and higher quality of service, from several perspectives (e.g. bandwidth; low latency); peering interconnections, as well as Cloud/CDN solutions, should be deployed wisely by taking into fair consideration the strategic and tactical implications of such technological configurations. Carriers' current financial configuration should be redesigned by evaluating the option of introducing innovative revenue sharing models and traffic charging schemes (e.g. Sending party pays model, Congestion charging) towards OTTs/CPs, with the ultimate goal of enhancing the business' financial sustainability.

These abovementioned considerations prove the reference model (solidly grounded in business model design theory) holds a straightforward value for Carriers' top managers, who can adopt it as a checklist of strategic and tactical decisions to be taken, and directly associate them to their implications.

Future studies should aim at testing the validity of the model within different firm samples, and, in turn, assess the relationship between specific choices in the business model parameters combination and firm performance.

Acknowledgments. The research leading to these results has received funding from the European's Community Seventh Framework Programme (FP7/2007-2013) under grant agreement n° 248567 for the 'ETICS' Project. Further information is available at www.ict-etics.eu. The authors would like to thank all members of the ETICS project for many valuable discussions and feedback.

References

1. Amit, R., Zott, C.: Value creation in e-business. Strategic Management Journal 22, 493–520 (2001)
2. Ballon, P.: Business modelling revisited: the configuration of control and value. Info. 9(5), 6–19 (2007)
3. Clark, D., Wroclawski, J., Sollins, K., Braden, R.: Tussle in Cyberspace: Defining Tomorrow's Internet. IEEE Transaction on Networking 13(3) (June 2005)
4. Ghezzi, A., Balocco, R., Rangone, A.: How to get Strategic Planning and Business Model Design wrong: the case of a Mobile Technology Provider. Strategic Change 19, 213–238 (2010), doi:10.1002/jsc.871
5. Hedman, J., Kalling, T.: The business model concept: theoretical underpinnings and empirical illustrations. European Journal of Information Systems 12(1), 49–59 (2003)
6. Johnson, M.W., Christensen, C.M., Kagermann, H.: Reinventing your business model. Harvard Business Review, 50–59 (December 2008)
7. Meredith, J.: Building operation management theory through case and field research. Journal of Operations Management 16, 441–454 (1998)
8. Morris, M., Schinderhutteb, M., Allen, J.: The entrepreneurs business model: toward a unified perspective. Journal of Business Research 58(6), 726–735 (2005)
9. Osterwalder, A.: The Business Model Ontology. A proposition in a design science approach. PhD thesis, École des Hautes Études Commerciales de l'Université de Lausanne (2004)
10. Rappa, M.: Business Models on the Web: Managing the digital enterprise. North Carolina State University (2001)
11. Tapascott, D., Ticoll, D., Lowy, A.: Digital Capital: Harnessing the Power of Business Webs. Harvard Business School Press, Boston (2000)
12. Timmers, P.: Business models for electronic commerce. Electronic Markets 8(2), 3–8 (1998)
13. Walsham, G.: Interpretive case-studies in IS research – nature and methods. European Journal of Information Systems 4(2), 74–81 (1995)
14. Weill, P., Vitale, M.: Place to Space: Migrating to E-Business Models. Harvard Business Press, Boston (2001)

15. Yin, R.: Case Study Research: Design and Methods. Sage Publishing, Thousand Oaks (2003)
16. Yu, C.-C.: An Integrated Framework of Business Models for Guiding Electronic Commerce Applications and Case Studies. In: Bauknecht, K., Madria, S.K., Pernul, G. (eds.) EC-Web 2001. LNCS, vol. 2115, pp. 111–120. Springer, Heidelberg (2001)
17. Zott, C., Amit, R.: Business model design: an activity system perspective. Long Range Planning 43(2-3), 216–226 (2010)
18. Zwickl, P., Reichl, P., Ghezzi, A.: On the Quantification of Value Networks: A Dependency Model for Interconnection Scenarios. In: Cohen, J., Maillé, P., Stiller, B. (eds.) ICQT 2011. LNCS, vol. 6995, pp. 63–74. Springer, Heidelberg (2011)

Reputation-Aware Learning
for SLA Negotiation[*]

Mohamed Lamine Lamali[1], Dominique Barth[2], and Johanne Cohen[2]

[1] Alcatel-Lucent Bell Labs, Route de Villejust, 91620 Nozay, France
mohamed_lamine.lamali@alcatel-lucent.com
[2] Lab. PRiSM, UMR8144, Université de Versailles
45, av. des Etas-Unis, 78035 Versailles Cedex, France
{dominique.barth,johanne.cohen}@prism.uvsq.fr

Abstract. Assuring Quality of Service (QoS) over multiple Network
Service Providers (NSPs) requires to negotiate QoS contracts as Service Level Agreements (SLAs) between NSPs. The goal of an NSP is to
maximize its revenues by selling as much as possible SLAs. However, provisioning too much SLAs might increase the risk of violations of the committed QoS thus impacting the NSP's reputation. In order to determine
the appropriate provisioning strategies, we propose to extend existing
solutions based on Reinforcement Learning with reputation-awareness
so that NSPs maximize their revenues.

1 Introduction

Internet applications (*e.g.* Gaming, videoconferencing, etc.) are more and more
demanding toward network resources. Some of them might require Quality of
Service (QoS) guarantees in order to enhance the end-user experience.

This suggests the Network Service Providers (NSPs) of the Internet to assure QoS and receive compensations accordingly. To support inter-NSP QoS for
every service, a Service Level Agreements (SLA) has to be committed between
a customer NSP (called *customers* in this paper) and one of its neighbor NSP
so as to build up an SLA chain to the destination. An SLA specifies possible
QoS guarantees (bandwidth, delay, etc.) and charging conditions (*e.g.* price for a
duration). For a given customer request of service, NSPs compete on their SLAs
and their prices. The customer will choose an SLA among its neighbor NSPs
according to its own utility (sensitivity to the proposed QoS, the price and reputation). In this competition, the provisioning strategy of an NSP is crucial as
it conditions the potential SLA violations. SLA violations play the main role in
the definition of the NSP reputation.

In this paper, we do not address the reputation propagation mechanisms
among customers. We aim to propose a model for NSPs to adapt their SLA

---

[*] This work has been partially supported by the ETICS-project, a project funded by
the European Commission through the 7th ICT-Framework Program. Grant agreement no.: FP7-248567 Contract Number: INFSO-ICT-248567.

Z. Becvar et al. (Eds.): NETWORKING 2012 Workshops, LNCS 7291, pp. 80–88, 2012.

provisioning strategy to meet customers' sensitivity to NSPs' reputation. We opt for a Reinforcement Learning approach as proposed by the authors of [5] and extend their model to take into account the NSP reputation.

Section 2 describes the competition among NSPs and the SLA negotiation problem. Section 3 describes the modeling of this problem. Section 4 presents the Reinforcement Learning algorithm and its using to tackle the problem of SLA negotiation. In sec. 5, we presents the results of simulation performed on an example of inter-NSP network.

2 The SLA Negotiation Problem and NSPs' Reputation

2.1 Competition among NSPs

The context of this paper is illustrated by fig. 1(a). There is a set of customers (Customer 1 and Customer 2 on the figure) requesting services with QoS guarantees to their neighbor NSPs (NSP 1, NSP 2 and NSP 3 on the figure).

A **QoS request** of a customer c is a 3-uple q_c s.t. $q_c = (l_c, d_c, b_c)$ of threshold values (packet-loss, delay and bandwidth respectively) chosen in a finite and discrete set, as for instance the QoS classes defined in [8]. When a customer sends such a request, it sends it simultaneously to all its neighbor NSPs. In fig. 1(a), Customer 1 sends QoS requests to NSP 1 and NSP 2. Customer 2 sends its QoS requests to NSP 2 and NSP 3. When receiving a request, each NSP chooses one SLA in its list of SLAs to make an offer to the customer which sent the request. An **SLA** q_i^j of an NSP i, where j is the SLA index, is a 3-uple (l_i^j, d_i^j, b_i^j). After receiving all the offers, the customer selects an SLA according to its utility function. Each NSP has a limited capacity. The higher is the part of used capacity, the more the NSP risks to violate its SLAs (*i.e.*, do not comply the committed QoS parameters). The reputation of an NSP is defined as the ratio of the number of its violated SLAs over the number of all its accepted SLAs. We assume that customers share the reputation of NSPs and that:

1. Each NSP has a set of predefined SLAs, and when receiving a request it chooses one of these SLAs to make an offer to the customer,
2. The NSPs know neither the utility function of the customers nor the SLAs proposed by the other NSPs,
3. The customers are sensitive to the price and the QoS parameters of an SLA, and also to the reputation of an NSP.

In such a context, the goal of an NSP is to maximize its revenues by selling its SLAs to the customers. The SLA proposed by an NSP must comply with the customer's QoS request and must not be too expensive in order to avoid customer's rejection. On the other hand, if an NSP sells too many SLAs (and thus uses much of its capacity), the probability of violation of its SLAs increases, therefore its reputation decreases and its next offers will be rejected because of its low reputation. Thus, each NSP must make a trade-off between selling its SLAs and keeping its used capacity low (and its reputation high).

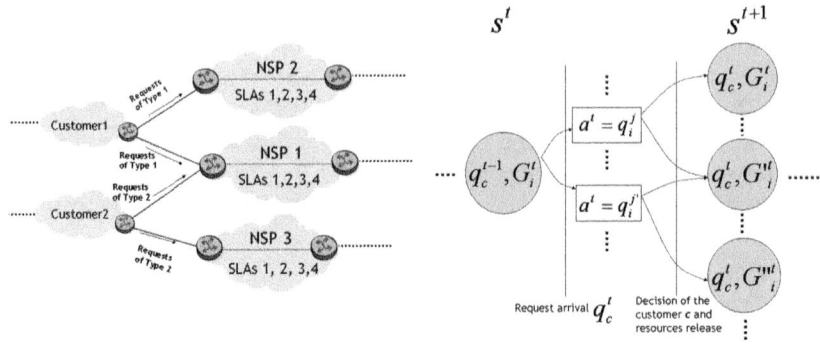

(a) Example of inter-NSP network   (b) Example of a part of an MDP

Fig. 1. Inter-NSP network and MDP

This paper aims to propose an algorithmic solution based on Reinforcement Learning in order to maximize the NSP's revenues by learning the best trade-off between sold SLAs and used capacity. Thus we do not address the issue of next hops and end-to-end computation.

2.2 Related Work

In [5] the authors propose two distributed algorithms based on Reinforcement Learning to compute an SLA chain insuring end-to-end QoS guarantees. The model of the customer acceptance presented in [5] is based on the price and the QoS but does not take into account the NSP reputation. The relation between the reputation and the provisioning strategies is also not considered. The authors of [3] propose a model where NSPs develop an economic alliance allowing them to share their knowledge and improve their revenues. This model takes into consideration the NSPs' capacity but not the reputation. The authors of [9] propose a path selection algorithm an provide a game theoretic analysis and equilibrium policies in the case of two NSPs.

The authors of [7] provide an analytical study of reputation based systems. Some recent works [1,2] propose management risk and ranking mechanisms to assure QoS in Grid Systems.

The goal of this work is a bit different as it proposes a model of competition taking into account the relation between capacity and reputation, and also the customers' sensitivity to the NSPs' reputation. This works also aims to provide mechanisms that allows NSPs to:

- Infer the customers' sensitivity to the reputation,
- Optimize their long-term revenues by finding a trade-off between selling SLAs and reputation,
- Answer in real-time to customers' requests,
- Take their decisions independently without cooperation (and thus avoid disclosing confidential information).

3   Modeling Inter-NSP Competition

This section describes the decision system of an NSP and how it can be modeled a Markov Decision Process (MDP). Reinforcement Learning Theory is based on MDP proprieties.

3.1   Decision System of an NSP

The remaining capacity of an NSP i is modeled as a *capacity grade* denoted G_i. It evolves according to the customers' reservations. To avoid the use of all possible values of the remaining capacity, the granularity of the capacity grade should be coarser in order to avoid combinatorial explosion of the number of states. Table 1 provides an example of possible values of G_i with corresponding levels of remaining capacity and probabilities of SLA violation. We consider a discretized time model where each step might not have the same duration but coincides with the decision taken at a request arrival. We assume that an NSP does not receive several requests at the same decision epoch t. The customers send simultaneously their requests to all their neighbor NSPs.

Table 1. Relation between capacity grade, remaining capacity and SLA violation

Capacity grade G_i	Remaining capacity	SLA violation probability
0	0% of total capacity	$f_0 = 1$
1	between 0% and 5% of total capacity	$f_1 = 0.95$
2	between 5% and 10% of total capacity	$f_2 = 0.8$
3	between 10% and 20% of total capacity	$f_3 = 0.6$
4	between 20% and 40% of total capacity	$f_4 = 0.2$
5	more than 40% of total capacity	$f_5 = 0.01$

An MDP is formalized as $\{\mathcal{S}, \mathcal{A}, P(.,.,.), \mathcal{R}(.,.,.)\}$ where $\mathcal{S}$ is the set of states, $\mathcal{A}$ the set of actions, the transition function $P(s, a, s')$ denotes the probability to move from a state s to a state s' when choosing an action a, and $\mathcal{R}(s, a, s')$ the reward obtained when choosing action a at a state s and moving to a state s'. According to our model, an MDP is defined for each NSP as follows:

- Each state $s \in \mathcal{S}$ is a pair (q_c, G_i) where q_c is a request of the customer c and G_i is a capacity grade. The state $s^t = (q_c^{t-1}, G_i^t)$ denotes the state at decision epoch t, q_c^{t-1} is a request treated at epoch $t - 1$, and G_i^t is the remaining capacity at epoch t. The state changes when a request is treated.
- The set of actions $\mathcal{A}$ corresponds to the set of SLAs of the NSP. At a decision epoch t an SLA is chosen, corresponding to action a^t, and proposed as an offer to the customer request. The chosen SLA must respect the QoS requirements of the request. Hence, $\mathcal{A}_t \subseteq \mathcal{A}$.
- The transition function $P(s, a, s') = \Pr(s^{t+1} = s' \mid s^t = s, a_t = a)$. As the last QoS request and the capacity state are included in each state, $P(s, a, s')$

is conditional upon q_c^t and G_i^{t+1}. The capacity grade G_i^{t+1} is itself conditional upon the release of network resources and the acceptance or refusal of the customer which issued q_c^t.

– If the chosen SLA $a^t = q_i^j$ is accepted by the customer, then the NSP reward $\mathcal{R}(s, a, s') = p_i^j$, the price of the SLA q_i^j. Otherwise $\mathcal{R}(s, a, s') = 0$. The reward obtained at decision epoch t is denoted $r^t = \mathcal{R}(s^t, a^t, s^{t+1})$.

3.2   Customer's Utility

As explained in sec. 2.1, the customers are sensitive to the QoS and the price of an SLA, but also to the NSPs' reputation. The utility of each customer increases according the QoS of the proposed SLA and to the reputation of the offering NSP, but it decreases according to the price of the SLA. Thus the utility function of a customer c associated to an SLA $q_i^j = (l_i^j, d_i^j, b_i^j)$ proposed by an NSP i is defined as:

$$U_c^t(q_i^j) = \frac{\rho_c ||q_i^j||_c + \eta_c \text{rep}^{t-1}(i)}{\beta_c p_i^j} \tag{1}$$

where:

– $||q_i^j||_c$ is a QoS measure of q_i^j and is equal to $\frac{l_c}{l_i^j} + \frac{d_c}{d_i^j} + \frac{b_i^j}{b_c}$,
– $\rho_c \in \Re_+$ is the *QoS rate* of the customer; *i.e.*, the weight of QoS measure in the customer's utility,
– $\beta_c \in \Re_+$ is the weight of the price of an SLA in the customer's utility,
– $\text{rep}^{t-1}(i)$ is the reputation of NSP i at decision epoch $t-1$, it is defined in sec. 3.3. The value η_c is the weight of reputation in the customer's utility.

3.3   Provisioning Strategies and Reputation

Remaining Capacity and Violations. As illustrated by Table 1, we consider that there is a strong relation between the level of available capacity and the probability of SLA violations (or trouble in general). In real networks, the level of remaining capacity is usually quite high to avoid troubles but allowing more resources to be provisioned might be in the interest of NSPs which can thus delay their investments and earn more revenues. Moreover, whatever the level of provisioned capacity is, the risk of trouble exists. The authors of [6] provide a tree of trouble causes in networks and their occurences.

Violations and Reputation. In our model, the reputation evolves according to the customers' experience. The SLA proposed by an NSP i can be violated with some probability denoted f_k. The customers take into account their own experience (how many times the SLAs proposed by some NSP failed) to determine the NSPs' reputation. The reputation of an NSP i is at decision epoch t is defined as:

$$\text{rep}^t(i) = 1 - \frac{\#fail(i)}{\#select(i)} \tag{2}$$

where $\#fail(i)$ is the number of times the offers of i were selected by a customer and was violated, and $\#select(i)$ is the number of times the offers of i were selected by a customer.

4   The Q-Learning Algorithm

We focus on the Q-learning algorithm because of its "model-free" ability (*i.e.*, it does not require a complete definition of function $P(s, a, s')$). Hence, it is particularly adapted to the inter-NSP SLA competition. The Q-Learning algorithm [10] learns optimal **Q-values** of each pair (state, action) at each decision epoch t. A Q-value of a pair (state, action) at decision epoch t is defined as $Q_t(s, a) = \mathbb{E}[R^t | s^t = s, a^t = a]$ where $R^t = \sum_{k=0}^{T} \gamma^k r^{t+k}$. The Q-values are updated at each iteration according to formula (3). These values converge to the expected gain corresponding to the definition above.

$$Q_{t+1}(s, a) \leftarrow (1 - \alpha^t)Q_t(s, a) + \alpha^t(r^t + \gamma \max_{a' \in \mathcal{A}} Q_t(s', a')) \qquad (3)$$

Algorithm 1 initializes the Q-value of each pair (state, action) and updates it when observing the reward r_t and using a discount factor γ and a "learning-rate" denoted α (α^t denotes the value of the learning-rate at decision epoch t). This latter also evolves at each decision epoch. A Q-based policy is the way to select an action based on Q-values. We focus on the ϵ-greedy policy because its behavior is adapted to the trade-off exploration/exploitation. The ϵ-greedy selects the action having the highest Q-value with a probability $1 - \epsilon$, and a random action with probability ϵ.

Convergence. The Q-Learning algorithm is proven to converge to optimal Q-values under two assumptions[4]: all pairs (state, action) must be visited infinitely, and $\sum_{t=0}^{\infty} \alpha^t = \infty$ and $\sum_{t=0}^{\infty} (\alpha^t)^2 < \infty$. There is an upper bound according to the update of α. If α is updated using a polynomial function ($\alpha^t = \frac{1}{(t+1)^\omega}$, with $\frac{1}{2} < \omega < 1$) then the convergence time is polynomial in $\frac{1}{1-\gamma}$. If $\omega = 1$ then the convergence time is exponential in $\frac{1}{1-\gamma}$.

Algorithm 1. Q-Learning algorithm

Initialization
loop
    At each decision epoch t
        Select an SLA $a^t = q_i^j$ according to ϵ-greedy policy
        Observe reward r^t and new state s^{t+1}
        Update the Q-values according to formula (3)
end loop

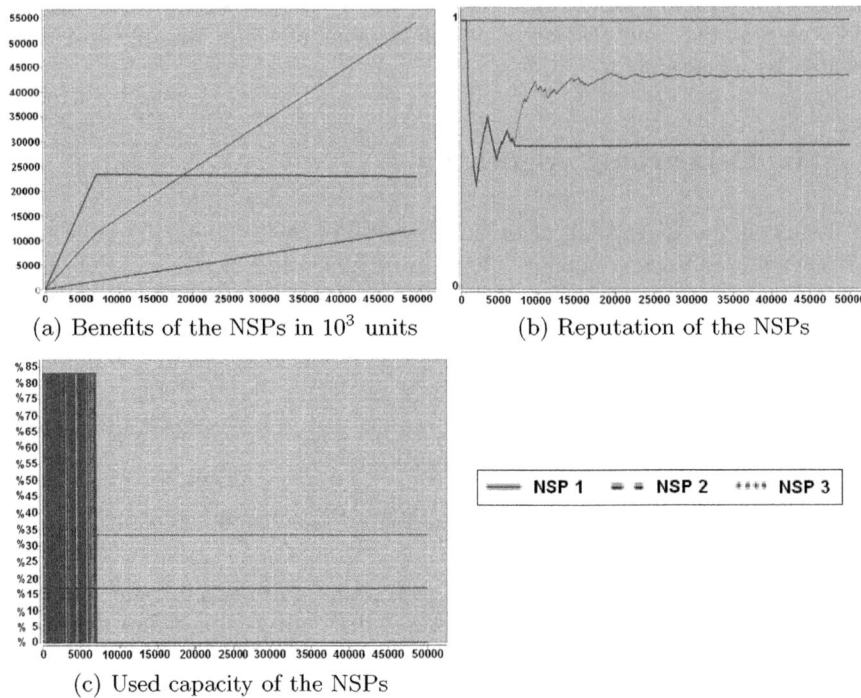

(a) Benefits of the NSPs in 10^3 units   (b) Reputation of the NSPs

NSP 1   ▪ ▪ NSP 2   ▪▪▪▪ NSP 3

(c) Used capacity of the NSPs

Fig. 2. Simulation results

5   Experiments

This section relates the results of the simulation performed on the network illustrated by fig. 1(a). Each NSP has a set of 4 SLAs. The first SLA $q_i^1 = (0.1\%, 30\text{ms}, 250\text{Mbps})$, the second one $q_i^2 = (0.05\%, 20\text{ms}, 500\text{Mbps})$, the third one $q_i^3 = (0.01\%, 10\text{ms}, 1000\text{Mbps})$ and the fourth one $q_i^4 = (0.001\%, 5\text{ms}, 2500\text{Mbps})$, for $i \in \{1, 2, 3\}$. According to a function of the SLA parameters, the price of the first SLA is $p_i^1 = 83$ units, the price of the second one is $p_i^2 = 250$ units, the price of the third one is $p_i^3 = 1000$ and the price of the fourth one is $p_i^4 = 5000$ units. Each NSP has a capacity of 3000Mbps. The first customer sends a request $q_c = (0.05\%, 20\text{ms}, 500\text{Mbps})$ at each decision epoch t. The request duration is 2 steps (if the request is sent at decision epoch t, the resource release occurs at epoch $t + 2$). The second customer sends a request $q_c' = (0.01\%, 10\text{ms}, 1000\text{Mbps})$ at each decision epoch t and its durations is also 2 steps. The utility functions of the two customers are the same, with $\rho_c = \rho_c' = 1$, $\eta_c = \eta_c' = 5$ and $\beta_c = \beta_c' = 1$. All the NSPs use the Q-Learning algorithm over their own MDP, as described in sec. 3.1. The violation probabilities are those indicated in Table 1.

Figure 5 shows the results of a simulation of 10^5 steps. Figure 2(a) shows that after a phase of learning of about 7000 steps, NSP 2 and NSP 3 monopolize the market and NSP 1's offers are never selected by the customers. Figure 2(b) shows that the reputations of NSP 1 and NSP 3 decrease during the learning phase, but after this phase NSP 3's reputation increases again and stabilizes at 0.8. The NSP 1's reputation stagnates at 0.55 after NSP 1 is never selected. Figure 2(c) shows that NSP 3's used capacity stabilizes at 33%, thus the remaining capacity stabilizes at 67%, which corresponds to a capacity grade $G_2 = 5$.

It appears that after a learning phase, NSP 2 and NSP 3 are in a dominant situation. NSP 1 uses must of its capacity because it offers its SLAs to both customers, and thus its reputation decreases and its offers are rejected. NSP 3 learns a trade-off between used capacity and SLA violation, and thus it recovered a better reputation when its used and remaining capacity stabilizes. Thus, it appears that if there is a dominant position for an NSP, the Q-Learning algorithm converges to the corresponding strategy and learns what is the amount of capacity which maximizes the NSP's revenues while keeping a good reputation.

6 Conclusion

In this paper, we address the SLA negotiation problem taking into account the SLA violation probabilities and their impact on the NSP reputation. The model presented in this paper takes into account the relation between the provisioned capacity of an NSP and the violation probabilities of its SLA. We propose a solution based on Reinforcement Learning techniques in order to allow NSPs to learn the best trade-off between provisioned capacity and good reputation. The simulations performed show that the proposed algorithm is able to learn this trade-off and maximize the NSPs' revenues.

As a future work, we plan to study other learning algorithms and the impact of the different learning parameters on the convergence. Another perspective is to investigate the possible existing equilibria between the NSPs.

References

1. Keller, A., et al.: Quality assurance of grid service provisioning by risk aware managing of resource failures. In: CRiSIS (2008)
2. Battré, D., Djemame, K., Gourlay, I., Hovestadt, M., Kao, O., Padgett, J., Voß, K., Warneke, D.: AssessGrid Strategies for Provider Ranking Mechanisms in Risk–Aware Grid Systems. In: Altmann, J., Neumann, D., Fahringer, T. (eds.) GECON 2008. LNCS, vol. 5206, pp. 226–233. Springer, Heidelberg (2008)
3. Barth, D., Mautor, T., Monteiro, D.V.: Impact of Alliances on End-to-End QoS Satisfaction in an Interdomain Network. In: ICC (2009)
4. Even-Dar, E., Mansour, Y.: Learning Rates for Q-Learning. Journal of Machine Learning Research 5 (2003)
5. Groléat, T., Pouyllau, H.: Distributed inter-domain SLA negotiation using Reinforcement Learning. In: Integrated Network Management (2011)
6. Medem, A., Akodjenou, M.-I., Teixeira, R.: Troubleminer: Mining network trouble tickets. In: IFIP/IEEE International Symposium on Integrated Network Management Workshops, IM 2009 (June 2009)

7. Nisan, N., Roughgarden, T., Tardos, E., Vazirani, V.V.: Algorithmic Game Theory. Cambridge University Press (2007)
8. Seitz, N.: ITU-T QoS standards for IP-based networks. IEEE Communications Magazine 41(6) (June 2003)
9. Suksomboon, K., Pongpaibool, P., Aswakul, C.: An equilibrium policy for providing end-to-end service level agreements in interdomain network. In: WCNC (2008)
10. Watkins, C.J.C.H., Dayan, P.: Technical Note Q-Learning. Machine Learning, 8 (1992)

From Quality of Experience to Willingness to Pay for Interconnection Service Quality*

Andreas Sackl, Patrick Zwickl, and Peter Reichl

FTW Telecommunications Research Center Vienna,
Tech Gate Vienna, Donau-City-Straße 1/3rd floor, 1220 Vienna, Austria
{sackl,zwickl,reichl}@ftw.at
http://www.ftw.at

Abstract. Triggered by the pending success of Quality of Service (QoS) differentiation in practice, recently the research interest has been increasingly focussing on the user-centric marketization of QoS, especially for interdomain scenarios. In this context, the key role of readiness and willingness to pay for enhanced network quality has not sufficiently been covered so far. In this paper, we focus on the users' willingness to pay for realtime network quality in interactive Video-on-Demand scenarios from an empirical perspective. Our user trial results indicate a broad willingness to pay for enhanced network quality, as well as remarkable influences on the quality perception through purchasing decisions, which is expected to kick off further experiments and the adaption of existing Quality of Experience models.

Keywords: Quality of Service, Quality of Experience, Willingness to Pay, Video on Demand.

1 Introduction

Quality of Service (QoS) differentiation for network services has been the center of many discussions in the past and presence, recently also in the interconnection (IC) context for providing quality guarantees beyond domain borders, e.g., in the ETICS project[1]. Especially in times of rapid network traffic growth rates in IC opposed to stagnating revenues, there is an increasing conviction that well configured QoS differentiation mechanisms may be economically advocated.

In practice however, the strong inherent needs of relating QoS to economics and charging policies – as often stated in literature (e.g., [1]) – and of aligning

---

* The authors would like to thank Sebastian Egger and Raimund Schatz for their support and fruitful discussions. The research leading to these results has received funding from the European Community's Seventh Framework Programme (FP7/2007-2013) under grant agreement n° 248567 for the ETICS project. Further information is available at www.ict-etics.eu. FTW is funded within the COMET Program by the Austrian Government and the City of Vienna.

[1] https://www.ict-etics.eu/

QoS differentiation mechanisms to customer demands, i.e., Quality of Experience, have been insufficiently satisfied by the available technical solutions, which have lead to the debatable success of network QoS so far [2].

This is further hampered by the diverse definitions and models available for QoE which have rather focused on integrating the user's perspective in the overall network environment [3,4,5,6,7,8] or quantifying the relationship between audio and video qualities and QoE [7,9,10]. However, these works have often failed to link the economic aspects in the users perceptions to concrete terms like the willingness to pay for certain QoS levels. Hence, this work contributes to fill this gap by investigating the end customers willingness to pay for improved network transmission quality by means of a user trial. According to [11], willingness to pay quantifies the amount of money to be spent for a given quality, while the related concept of readiness to pay may be defined as the general disposition to pay for a given quality. Positioned between these two bounds, our empirical study investigates how much end customers pay for Video-on-Demand (VoD) streaming quality in order to better integrate end customers interest in quality differentiations.

Related work on this topic is rather limited. Probably closest to our work, the study [12] has initially linked the idea of different quality opportunities with consumers's monetary decisions. In contrast to our approach, [12]focused on pre-rendered video qualities, i.e., videos were presented with different bitrates. The monetary decisions were analyzed by randomly assigning users to user profiles with varying prices for predefined video bitrates. Within this setup, the M3I experiments demonstrated users willingness to pay for higher QoS. Based on that, our research intends at transferring these monetary aspects of video qualities to realtime transmission interconnection scenarios, i.e., a common VoD-scenario, where QoS is modeled in terms of different packet loss rates and linked to a certain pricing structure, in order to identify implications on the technical management of network QoS classes and their marketization in terms of sellable goods.

The contribution of the present work is therefore twofold: on the one hand we investigate the willingness to pay for improved network quality in a realistic real time video streaming scenario, and on the other hand we incorporate Quality of Experience aspects like acceptance rates and subjective video quality perceptions.

The rest of this paper is structured as follows: we start with describing in detail the technical setup and the procedure of the user trial. Section 3 discusses the results of our empirical study including user demographics, user behavior and of course outcomes regarding willingness to pay. Finally, Section 4 presents our conclusions.

2 Experimental Setup

In order to provide for an utmost realistic live streaming VoD scenario, we have taken recourse to FTW's iLab user test laboratory[2] for creating the atmosphere of

---

[2] FTW Interfaces & Interaction Lab (i:lab): http://www.ftw.at/portfolio/i-lab?
set_language=en, last accessed at Jan 26, 2012

a living room-like situation. Based on different levels of random packet loss rates, after extensive pre-tests, we have formulated four relevant network quality levels (i.e., price and packet loss percentage – see Table 1). This setup is in addition backed by an extensive library of modern movies, TV series, and documentations, as well as by realtime UDP network transmission of the movies with a realistic constant delay of *75 ms* and a variable packet loss on the used link.

Table 1. The offered quality classes

	Quality 1	Quality 2	Quality 3	Quality 4
Packet loss [%]	1	0.25	0.085	0
Delay [ms]	75	75	75	75
Costs per min [€]	0	0.025	0.05	0.075
Costs per movie [€]	0	0.5	1	1.5

Procedure. Inspired by [12], our experiment creates an interactive quality market allowing the consumers to purchase realtime quality enhancements of the streamed video with real money. For this purpose, each user is assigned a balance with an initial deposit of *€10* for their free disposal in the experiment and/or afterwards. Thereafter, the users individually watch three videos of their choice with a duration of *20 minutes* each, which are automatically launched in the worst available quality. Each movie starts with an initial *trial phase* (2-3 minutes duration, free of charge) for experimenting with the different quality levels and eventually deciding (and purchasing) one quality level which is provided afterwards. During the video runtime, this pattern is repeated three more times, i.e., trial phases for upgrading the original decision are also offered after *5, 10* and *15* mins (note that no downgrading is allowed throughout). At the end of the experiment the remaining deposit (Euro) is paid out in cash to the users hence capturing the willingness to invest the own money for network quality upgrades.

Laboratory. From the technical perspective, in our setting the video stream is triggered by an *iPad* (cf. Figure 1). The test users can select a video of their interest from our extensive collection. By choosing a video, the iPad calls a script which starts the VLC[3] video stream on our *Linux server*. At the same time, a packet loss script randomly dropping packages at a given percentage is initiated through a netem[4] command providing network emulation capabilities for Linux systems. This sets up the defined network quality for the starting phase – the worst quality offered in our experiment. The video is then streamed via the network to a *Mac Mini Server*, which displays the received videos on a directly connected *flat screen television*. The Mac Mini in this case appears to the user as set-top-box, hiding the details of the transmission logic, i.e., network details

---

[3] VideoLAN Client (VLC): http://www.videolan.org, last accessed at Jan 26, 2012.
[4] Netem: http://www.linuxfoundation.org/collaborate/workgroups/networking/netem, last accessed at Jan 24, 2012

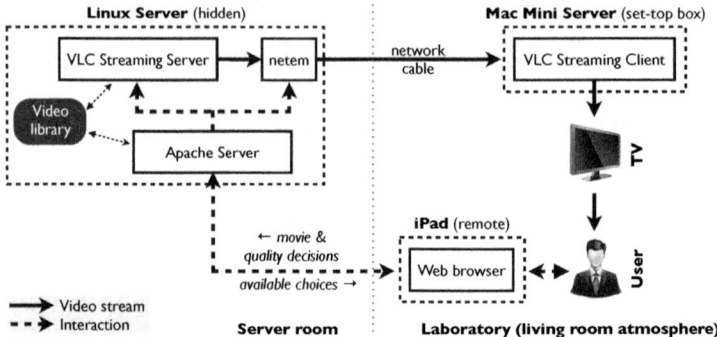

Fig. 1. Technical setup of the experiment

and server structures. The iPad, moreover, acts as remote control offering the users to purchase quality upgrades.

Our technical setup also foresees the capability of constantly increasing the packet loss between two *trial phases* without explicitly notifying the test users. This process is applied between the first and the second trial phase of the third movie. The packet loss is continuously increased by *0.2%* in a modification interval of *60 seconds*, while leaving the price calculation stationary. The resulting packet loss deterioration then remains active for the residual part of this movie and is cleared afterwards again. With these data, we intend to gain further understanding on relevant triggers for the users' quality decisions.

3   Results

Based on the concept described above, a laboratory-based user trial was scheduled in Vienna in October 2011. Its results have been intensively statistically analyzed and are presented in this section.

3.1   Test User Demographics and Background

Overall, *43 users* (*22 male* and *21 female*) successfully participated in our study. Approximately *40%* were between *18* and *29* years old, *32%* were between *30* and *44* years old and *28%* were older than 45 – the mean age was *36.8*. Most of our users where employed (*48%*)or students (*28%*). More than *93%* of the test users were familiar with YouTube, *75%* of them used this service at least once a week. Most of them have consumed music videos (*67%*), while movies and fun videos have been of limited interest (*20%*). On the other hand, only *5* users have consumed videos from dedicated VoD platforms (*2x* iTunes, *2x* A1 Videostore, *1x* UPC on demand) with a mean monthly spending of *€5.48* before.

3.2   Willingness to Pay

Figure 2 depicts each users remaining deposit after the consumption of the three videos. A subset of *9* "generous" users have spent the maximum amount of money (€*4.5* of their €*10* balance i.e., €*1.5* per movie), while *4* "budget-minded" users decided to watch all three movies in the worst quality (and receiving the maximal payout of €*10* in cash). The majority, however, has taken an intermediary position between these two extremes.

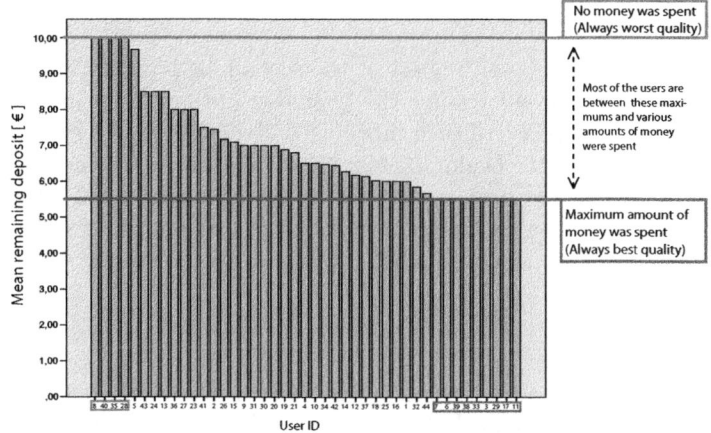

Fig. 2. Users' remaining deposits after three rounds

The average of all 43 users has spent €*1.01* (±*0.49*, standard derivation) per movie to increase the quality, i.e., to decrease the packet loss of the transmission – corresponding to two-thirds of the maximum potential spending of €*1.5*. Consequently, this provides a proof for the feasibility of our setup linking real cash payouts with an laboratory setup. The spent money is obviously subject to the presented quality. The mean chosen quality level per movie was *3.04* (±*0.99*), i.e., on average the second best quality level was chosen. Most purchases have been made during the first trial phase of each movie. During the rest of the movie, only limited quality modifications have been applied. Not even continuously decreasing prices – as calculated on the basis of the remaining share of the movie – have triggered a higher intensity quality upgrades.

3.3   User Behaviors

The observed behavior of our participants provided a revealing insight in the chosen purchasing and consumption strategies:

– **Strategic players**: Some users have repeatedly chosen the best quality during the trial phases (free of charge). However before the end of each trial phase, they have returned to the poorest quality level again – i.e., no quality purchases have been made in order to receive the maximum deposit payout.

- **Generous players**: We could observe users always selecting the best quality at the beginning of the movie without testing lower qualities. Thereafter, they have stated their insensitivity to payments from their deposit and their intention to watch all movies in the best available quality.
- **Budget-minded players**: Some users have declared after the trial that they would not care about the quality as long as they would receive the full €10 payout. In some cases, better quality has not even been tested.
- **Quality & price-aware players**: Most of the users have tested all quality levels at the beginning and have finally chosen quality 3 or 4 (willing to pay the corresponding charges).

Generally, most users chose the best or the second best quality in the first trial phase without extensively testing the available options. The mean click rate during the first trial phase of each movie is *4.17* clicks, i.e., every user has only changed the quality at the beginning four times on average. During the rest of the movie, the click-rate has been even lower (mean click rate is *6.5*): the majority of users changed the quality less than *10* times per movie. The average click-rate per user remained constant over time respectively over the three selected videos.

3.4 Acceptance

Another interesting question concerns the relationship of perceived quality and user acceptance. Related work has e.g., studied the correlation between MOS ratings and acceptance, for example [13] examined this correlation for mobile broadband usage. There are also studies (e.g., [14]) that have examined the influence of codec settings, content types, various devices etc. on the acceptance. The analysis of different video qualities regarding MOS ratings and acceptability, however, to the best of our knowledge has not been addressed by the research community. Therefore, we have asked the test persons also to fill in a questionnaire on their perceived video quality after each video, i.e., ACR 5 MOS-scale [15], and whether they would chose such a quality at home, i.e., a video acceptance rate. To our surprise, even lower video quality levels are acceptable for the majority of the users (Figure 3). These findings are currently object of investigation and further studies are planned to analyze this attitude.

3.5 Further Findings

There were no significant statistical correlations between the click-rate, the spent money or the chosen quality level and age, sex, education or YouTube usage of the participants. During the third movie, a hidden packet loss boost (*+0.2%*) has been initiated after the first trial phase for all quality levels lower than *4* (*31* of *43* test users did not choose the best quality level). While *21* users have not reacted to this hidden deterioration, *9* participants increased the quality by one level and one user increased it by two levels – the packet loss boost has only affected about one third of all eligible users.

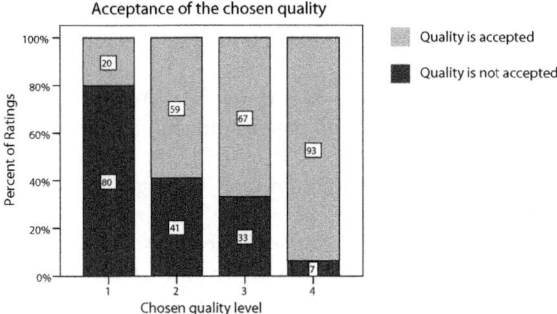

Fig. 3. Users' acceptance per qualityclass (packetloss). 1=worst quality; 2=best quality

4   Conclusions

Our analysis has demonstrated the feasibility of the used laboratory setup for realistic willingness to pay studies for VoD scenarios. This is also confirmed by the existence of a substantial willingness to spend money for enhancing the network quality, i.e., reducing network packet loss. Consequently, our work may serve as basis for more complex investigations regarding e.g., users' price elasticity to quality changes.

One particular phenomenon could be observed regarding the users interaction habits with the quality market. After their initial choice, a strong tendency of rarely modifying chosen quality levels has often educed the total ignorance of three out of four upgrade opportunities. We could also show the presence of various interesting purchasing strategies such as users continuously purchasing the highest quality or the lowest price option – anticipating four types of players.

Due to the limited user reactions (*33%*) on a hidden packet loss boost (*+0.2%*), we may argue that pricing has been predominately the decisive factor for purchasing decisions after the initial quality level choice. Like with a couple of other issues already mentioned earlier, further clarification on this point will be subject of future work.

References

1. Jain, R.: Internet 3.0: Ten problems with current Internet architecture and solutions for the next generation. In: Proceedings of the Military Communications Conference (MILCOM 2006), pp. 1–9. IEEE (2006)
2. Burghstahler, L., Dolzer, K., Hauser, C., Jähnert, J., Junghans, S., Macián, C., Payer, W.: Beyond technology: the missing pieces for QoS success. In: Proceedings of the ACM SIGCOMM Workshop on Revisiting IP QoS (RIPQoS 2003). ACM (2003)
3. Kilkki, K.: Quality of Experience in Communications Ecosystems. Journal of Universal Computer Science, Special issue on Socio Economic Aspects of Next Generation Internet, 615–624 (2008)

4. International Telecommunication Untion (ITU): Vocabulary for performance and quality of service, amendment 2: new definitions for inclusion in Recommendation P.10/G.100 (2008)
5. Lopez, D., Gonzalez, F., Bellido, L., Alonso, A.: Adaptive multimedia streaming over IP based on customer oriented metrics. In: International Symposium on Computer Networks (2006)
6. Soldani, D., Li, M., Cuny, R.: QoS and QoE Management in UMTS Cellular Systems. Wiley (2006)
7. Reichl, P., Tuffin, B., Schatz, R.: Logarithmic Laws in Service Quality Perception: Where Microeconomics Meets Psychophysics and Quality of Experience. Telecommunication Systems (2011)
8. Fiedler, M., Kilkki, K., Reichl, P.: From Quality of Service to Quality of Experience. In: Proceedings of the Dagsuhl Seminar 09192 (2009)
9. Staelens, N., Moens, S., Van den Broeck, W., Mariën, I., Vermeulen, B., Lambert, P., Van de Walle, R., Demeester, P.: Assessing Quality of Experience of IPTV and Video on Demand Services in Real Life Environments. IEEE Transactions on Broadcasting 56(4), 458–466 (2010)
10. Zhongkang, L., Weis, L., Boon Choong, S., Kato, S., Eeping, O., Susu, Y.: Perceptual Quality Evaluation on Periodic Frame Dropping Video. In: Proceedings of the International Conference on Image Processing. IEEE (2007)
11. Ries, M., Nemethova, O., Rupp, M.: On the Willingness to Pay in Relation to Delivered Quality of Mobile Video Streaming. In: Proceedings of the International Conference on Consumer Electronics. IEEE (2008)
12. Hands, D. (ed.): FP5 Project M3I, IST1999 11429, Deliverable 15/2 – M3I user experiment results (2002)
13. Schatz, R., Egger, S., Platzer, A.: Poor, Good Enough or Even Better? Bridging the Gap between Acceptability and QoE of Mobile Broadband Data Services. In: Proceedings International Conference on Communications (ICC 2011). IEEE (2011)
14. Knoche, H., McCarthy, J., Sasse, M.: Can small be beautiful?: assessing image resolution requirements for mobile TV. In: Proceedigs of the 13th International Conference on Multimedia (2005)
15. International Telecommunication Untion (ITU): ITU-R BT.500-7 Methodology for the subjective assessment of the quality of television pictures (2002)

An Instance-Based Approach for the Quantitative Assessment of Key Value Network Dependencies[*]

Patrick Zwickl[1] and Peter Reichl[1,2,3]

[1] FTW Telecommunications Research Center Vienna, Austria
{zwickl,reichl}@ftw.at
[2] CNRS, LAAS, 7 avenue du colonel Roche, F-31077 Toulouse Cedex 4, France
[3] Université de Toulouse; UPS, INSA, INP, ISAE; UT1, UTM, LAAS;
F-31077 Toulouse, Cedex 4, France

Abstract. Entailed from challenges in industries which are characterized by competitive economics of scale, like for instance network interconnection, the interest in Value Networks (VN) has significantly grown recently. While most of the related work is focussing on qualitative VN analysis, in this paper we describe an enhanced model of a VN quantification concept and argue for the instance-based orthogonalization of key VN dependency indicators. As a result, the reduced set of such independent indicators is organized around the axes of customers and suppliers in order to form manageable quantification algorithms.

Keywords: Value Networks, Quantification, Dependency Analysis, Fator Analysis, Interconnection.

1 Introduction

The economic challenges in highly competitive industries such as network Interconnection (IC)—stagnating revenues[1] being opposed by tremendous demand growth rates[2]—require conceptual (non-linear) inter-firm supplements to Business Models [1,2], i.e., *Value Networks* (VNs) [3,4,5]. However, related work focuses on qualitative mechanisms for comparing and analyzing available VN options such as [6], while to the best of our knowledge, hardly any approach providing quantitative support for assessing VN options has been proposed so far.

---

[*] The research leading to these results has received funding from the European Community's Seventh Framework Programme (FP7/2007-2013) under grant agreement n ° 248567 for the ETICS project. FTW is funded within the COMET Program by the Austrian Government and the City of Vienna.
[1] DrPeering International—Internet Transit Prices – Historical and Projected:
http://drpeering.net/white-papers/Internet-Transit-Pricing-Historical
-And-Projected.php, last accessed: Jan 31, 2012.
[2] Cisco Visual Networking Index Forecast Projects 26-Fold Growth in Global Mobile Data Traffic From 2010 to 2015: http://newsroom.cisco.com/press-release
-content?type=webcontent&articleId=5892556, last accessed: Jan 30, 2012.

Z. Becvar et al. (Eds.): NETWORKING 2012 Workshops, LNCS 7291, pp. 97–104, 2012.
© IFIP International Federation for Information Processing 2012

Therefore in the compagnion paper [7], we have introduced a quantified dependency quantification method for VNs. With this concept, the dependency of an entity e on the VN, i.e., on all other entities, can be calculated on the basis of six relative dependency indicators (mainly derived from Porter's five forces on firms [8]): Bargaining power of suppliers (δ_1^e) and customers (δ_2^e), substitutes (δ_3^e), potential market entrants (δ_4^e), industry rivalry (δ_5^e), resource type dependency—i.e., fungibility of resources (δ_6^e). These indicators are supposed to measure forces in VNs which might cause structural changes of businesses and related VNs in the future. The six dependency indicators δ_i^e are aggregated to form a single dependency factor Δ^e,

$$\Delta^e = \sum_{i=1}^{6} w_i * \delta_i^e \quad , \tag{1}$$

for entity e by the usage of weighting factors w_i. While the dependencies themselves are extracted from entity-specific data, e.g., price and cost values (for details we refer to [7]), we have abstained from discussing the role of the weighting factors in detail so far.

By applying a more fine-granular instance-based perspective, in this paper we aim at rendering several of these weighting factors obsolete (dependency from substitutes, potential entrants, and industry rivalry—see Figure 1), while we consider the orthogonalization of key VN dependency indicators and the explicit modelling of substitution to be additional novel contributions of this work. For this purpose, we extend the concept described in [7] by revisiting the dependency indicators $\delta_{\{3,4,5,6\}}^e$ while $\delta_{\{1,2\}}^e$ are left untouched. For further details on our VN

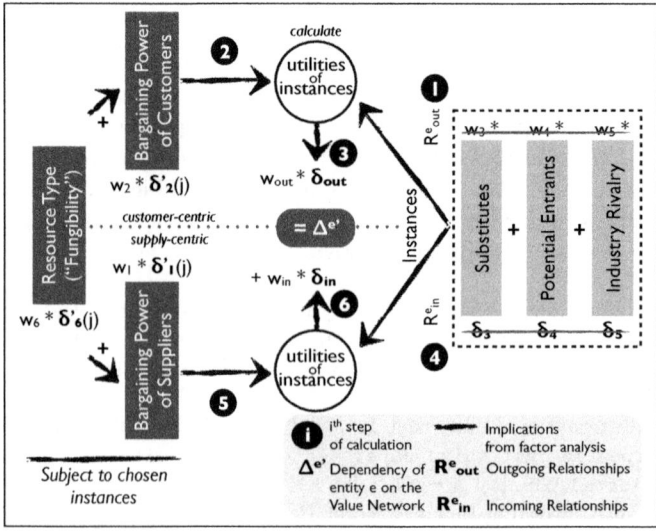

Fig. 1. Orthogonalization of initial dependency indicators from [7]

quantification concept, the reader is strongly refered to [7] as starting point for the subsequent discussion and extension.

The remainder of this paper is structured as follows: We first identify required VN quantification assumptions in Section 2, which are then used as conceptual basis for an instance-based VN dependency indicator orthogonalization concept in Section 3. In Section 4, an evaluation discussing potential interpretations and limitations is given. We conclude in Section 5 with a summary of key findings.

2 Assumptions

For the construction of our fine-granular entity dependency model, we apply a series of assumptions and interpretations followed throughout the work.

Monopoly. Our model relies on the monopolistic decisioning case fully eliminating resource scarcity and competition for resources. Hence, all alternatives are assumed to be available when bearing a certain investment effort. Competitive games for resources may have to be considered in future work.

Stationarity. The core of the quantified VN options are assumed to remain stationary throughout the analysis, i.e., prices, costs, players (entities and their instances), relationships, substitutes, investments costs etc. remain constant. Hence, all substitution options can be explicitly modeled a priori.

Dynamicity. The dynamicity of real-world VNs is captured through modeling player entrance and potential player exits, as well as the analysis of substitutes.

Utility. Added values are used in the sense of utility gains—based on the Von Neumann-Morgenstern utility theorem [9]—already incorporating values, costs, risks, chances, and further impairments.

Time. Utilities are standardized for a predefined *time span* with a clearly defined *starting time*. This is required in order to sufficiently integrate investment costs that may be required in order to switch to a substitutive alternative. This concept strongly relates to depreciation mechanisms used in accounting.

Complete Knowledge. The information required in order to allow a sufficient dependency reasoning process (complete knowledge) is taken as given.

Individual Rationality. We assume individual rationality of demanding customers (i.e., only positive utilities for buying a product may trigger a purchase) and involved entities, i.e., always the individually best VN option is chosen.

Entity. The identification of the best VN configuration is subject to the perspective of an entity or a group of entities minimizing its dependency towards the VN.

3   New Model Features

The present work further details the VN dependency quantification as proposed in [7] by orthogonalizing the key VN dependency indicators. So far, the dependencies $\delta^e_{\{3,4,5\}}$ resulting from industry rivalry, substitutes, and potential market entrance have been treated as individual and mutually independent indicators. In order to make them quantitatively comparable in a unified representation, we transfer the entity-based perspective of VN quantification to a finer-granular instance-based analysis. Basically, we consider each individual relationship instance as an *option* for providing a certain resource (due to either industry rivalry, substitution or market entrance), and calculate its utility. In this way, the dependency of one instance from all other (instance-based) alternatives is quantified. To this purpose, the representation of VNs is slightly enhanced as basis for further specifying the calculation of VN dependencies.

3.1   Substitution and Market Entrance

So far, the Value Network Dependency Model (VNDM) [7]—an approach for visually sketching VNs—has not captured substitutes. In principle, *substitutes* may be represented in form of separate VNDMs. For a better overview, we however have opted for directly integrating them in a single VNDM.

Visualization mechanisms immanently face a trade-off between their complexity and their expression power. Thus, the set of relationship alternatives are represented by the most beneficial (utility maximizing) instance choice—especially also integrating instances from potential *market entrance*—for the VN (in terms of resource values, costs, etc.). Each available substitution is visually marked in the model in order to reflect required investments.

In the example of Figure 2, a simple Video-on-Demand (VoD) interconnection (IC) scenario is sketched (an excerpt of the example used in Figure 3 in [7]). A movie platform from New York is delivering a movie to an end customer's television set in London. For this purpose, the platform provider pays for the Best-Effort delivery ($1.5) in order to receive a delivery promise from its access

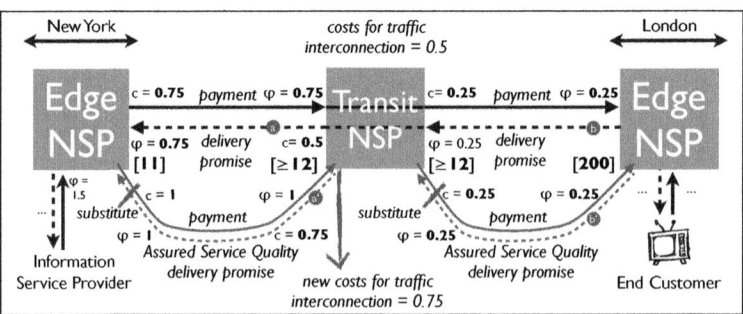

Fig. 2. Partial view on a Video-on-Demand VNDM ([7] Fig. 3) extended with ASQ traffic delivery as substitute

provider, i.e., an Edge Networks Service Provider (NSP). This NSP (similar to [7] we assume there are 11 Edge NSP instances) is then responsible for establishing the IC via a Transit NSP ($\geq$ 12 instances [7]) to the Edge NSP (200 instances) delivering the content to the customer, i.e., it pays the chain of Transit and Edge NSP (cost $c = \$1$). In addition, we have added in Figure 2 Assured Service Quality (ASQ) traffic interconnection as substitute for the chosen Best-Effort solution. For ASQ, we assume higher costs c (labelled at the source of the edge), but also higher revenues/value φ (at the tip of the arrow).

3.2   Dependencies from Alternatives

The revised entity dependency calculation proposed in this section addresses two dimensions: (i) the available relationships, and (ii) the utility of available instances for each relationship. These dimensions are targeted by forming a detailed utility model, which incorporates dependencies from inter-relationship factors. We further distinguish customer-side and supply-side dependencies.

Customer-side Dependencies. For the purpose of quantifying the customer-side dependencies, we enumerate and order the available instances for each entity (competition), market entrance options, and available substitutes by incorporating the required investment and other effort compensations in the utility assessment of an option (adjusted to the chosen time span). We call each of the corresponding alternatives an *option*, and will subsequently use these options to analyze the available alternatives for each relationship of a VN or its VNDM representation—i.e., the set of options $O(r)$ for each relationship r of an entity. Methodologically, we are taking recourse to the existing dependency calculation approach based on the *Gini coefficient* [10] as introduced in [7], while at the same time we aim at the harmonization of information on the stated forces.

In our case, the Gini coefficient indicates the fairness of utility distribution among available options for instantiating a relationship, i.e., other entities providing the same good or technical alternatives. Unfortunately, this parameter does not consider whether the second best options is dramatically worse than the optimum etc. This, however, is a requirement for assessing the dependency of one relationship (and its entity) on the availability of one particular instance. We assume that the utility divergence of options—ordered by their utilities—is intensified by an exponential spreading factor, i.e., the utility decreases from one option to the next-best one by a constant multiplicative factor < 1. Therefore, we define the exponential spreading factor $y(j)$ for the utility calculation, where $o(j)$ is the jth best relationship option of all n_r available options of relationship r of an entity e and $y(j)$ continuously decreases with growing j:

$$y(j) := \frac{1}{n_r} * e^{\frac{ln(n_r)}{n_r-1}*(n_r-j)} \qquad (2)$$

where $y(j) \in [1/n_r, 1], j \in \{1, 2, ..., n_r\}$, and $n_r = |O(r)|$.

Note that $y(1) = 1$ and $y(n_r) = \frac{1}{n_r}$, hence these marginal values match those obtained for instance also by a model of linearly decreasing utilities.

In order to make the available options quantitatively comparable, impairments such as necessary investment costs i or inherent risks—e.g., the risk of weak customer relationships—need to be incorporated in the utility calculation of options. By taking depreciation during the calculation of the utilities into account (adjusted to the chosen time span) their utilities can be directly compared, i.e., additional weighting factors are no longer required. In addition, the integration of alternative options (substitution, market entrance) entails the possibility of changing resource types, values or costs, hence rendering each option conditional to dependencies from resource types δ_6^e and the bargaining power of customers δ_2^e. The vague description of δ_6^e given in [7] is further detailed in Equation (3) for the jth best option. Explicitly, an exchanged resource may be used for a subset $k(j)$ of all possible business relationships M of the VN, $0 \leq k(j) \leq |M|$. The higher $k(j)$, the lower is the dependency on the VN—and vice versa.

$$\delta_6^{\prime e}(j) := 1 - \frac{k(j)}{|M|} \tag{3}$$

The overall utility $\mathcal{U}(j)$ incorporating all relevant dependency indicators for the jth best option $o(j)$ is calculated according to (4):

$$\mathcal{U}(j) := y(j) * [\varphi(o(j)) - c(o(j)) - i(o(j))]* \\ * w_{2'} * [1 - \delta_2^e(j)] * w_{6'} * [1 - \delta_6^e(j)] \quad , \tag{4}$$

where $\varphi(o(j))$ is the value, $c(o(j))$ is the operation cost, and $i(o(j))$ the investment cost (optional) of the jth best option for relationship r. In contrast to $w_{\{2,6\}}$ in (1), the revised weights $w_{\{2,6\}}'$ are integrated in the calculation of $\mathcal{U}(j)$ for a chosen j in (4). Hence, for every j both forces are calculated beforehand (step 2—cf. Figure 1), as if the jth option would have been the used (best available) variant—for all other relationships the best option is kept stationary. Thereafter, δ_{out}^e can be computed, which may imply a disproportional complexity increase with the number of relationship options. By using the utility calculation of (4), the adapted dependency calculation can be formulated as follows:

$$p(j) := \frac{\mathcal{U}(j)}{\sum\limits_{k \in \{1,\dots,n_r\}} \mathcal{U}(k)} \quad , \tag{5}$$

$$gini(r) := \sum\limits_{j \in \{1,\dots,n_r\}} [p(j)]^2 \quad , \tag{6}$$

$$gini(R^{e,out}) := \frac{1}{|R^{e,out}|} * \sum\limits_{r \in R^{e,out}} gini(r) \quad , \tag{7}$$

$$\delta_{out}^e := \frac{gini(R^{e,out})}{\max\limits_{k \in E} \{gini(R^{k,out})\}} \quad , \tag{8}$$

where $e \in E; r \in R^{e,out} \subseteq R^e \subset R$, and $gini(R^{e,out}) \leq 1$.

(5) and (6) calculate the Gini coefficient for each option of all available options $O(r)$ for a relationship r. This captures whether the utilities $\mathcal{U}$ are evenly distributed over alternatives, i.e., measuring whether the entity requiring this relationship has valuable options. Accordingly, the dependency of each instance increases with every valuable option to be replaced. In (7), the average Gini factor of all outgoing relationships $R^{e,out}$ of entity e is calculated. This is finally turned to a relative indicator for entity e's dependency on the VN in (8)—in respect to the highest dependency measured in the VN.

Supply-side Dependencies. The dependencies δ_{in}^e resulting from incoming edges (supply) are quantified in analogy to the customer side—see (8). However, $\delta_2(j)$ is replaced by $\delta_1(j)$ and in lieu of outgoing edges $R^{e,out}$ the incoming edges $R^{e,in}$ are used. Moreover, the dependency calculation is essentially modified in (9), i.e., the dependency decreases with the number of supplier options:

$$\delta_{in}^e := 1 - \frac{gini(R^{e,in})}{\max_{k \in E}\{gini(R^{k,in})\}} \quad . \tag{9}$$

Together δ_{in}^e and δ_{out}^e capture the dependency of entity e on the VN, where weighting factors $w_{\{in,out\}}$ with a sum of 1 are used for aligning the indicators, thus eventually replacing the equation of (1):

$$\Delta'^e := w_{in} * \delta_{in}^e + w_{out} * \delta_{out}^e \quad . \tag{10}$$

4   Interpretations and Limitations

The instance-centric perspective on VN dependencies has led to the formation of two key VN dependency factors eliminating the need to autonomously quantify dependencies from substitutes, market entrance, and industry rivalry. We suggest to denote these new indicators as customer-side and supply-side *dependencies from alternatives*. This modification reduces the complexity for the dependency calculation in practice. On the other hand, the concentration on an instance-based analysis has emphasized the disproportionally growing calculation effort with an increasing number of instances. In addition, the requirement of numeric market data estimations has been inferred from the quantitative methodology.

Rearrangements. Whenever a VN is extensively rearranged, the mapping of one relationship to a successor may require some interpretation. In particular, relationships may be logically merged or costs assigned to several relationships. Precise computational routines for such situations are left for further work.

Dependences. The almost unlimited fungibility of money [11] entails the lowest possible resource type dependency of the resource exchange on the VN—as captured in the adapted calculation of δ_6' (see Equation (3)).

Completeness. Although market entrance is captured explicitly in the model, the risk of market exits decreasing the utility of one relationship has only implicitly addressed as further risk. By graph analysis the completeness of the used dependency forces needs to be further confirmed.

5   Conclusions

The present work conceptually argues for the orthogonalization of key indicators for VN dependencies. The initial indicators have been derived from [7], where we have proposed to turn Porter's five forces into quantitative VN dependency indicators, i.e., metrics stating how dependent one entity is on the overall VN. In order to capture the interrelations between these initial indicators, we have formed two key dependency indicators δ_{in}^e (supply-side dimension) and δ_{out}^e (customer-side dimension) as a result of an instance-based analysis. This outcome has rendered several of the initial indicators redundant, which may reduce the complexity of VN dependency quantifications. The indicators δ_{in}^e and δ_{out}^e have been formed around the instance-based dependency arising from the threat of being replaced by another instance (whether through industry rivalry, substitution, or market entrance). Future work will target the confirmation of the initial dependency indicators providing a basis for the advanced dependency quantification Δ'^e shown in (10), as well as the practical quantitative assessment of VN alternatives, e.g., enabling ASQ traffic as depicted in Figure 2.

References

1. Timmers, P.: Business Models for Electronic Commerce. Electronic Markets 8(2) (1998)
2. Teece, D.J.: Business Models, Business Strategy and Innovation. Long Range Planning 43(2-3), 172–194 (2010)
3. Hakansson, H., Snehota, I.: No Business is an Island: The Network Concept of Business Strategy. Scandinavian Journal of Management, 187–200 (1989)
4. Normann, R., Ramirez, R.: Designing Interactive Strategy: From the Value Chain to the Value Constellation. John Wiley & Sons (1994)
5. Gulati, R., Nohria, N., Zaheer, A.: Strategic Networks. Strategic Management Journal 21, 203–215 (2000)
6. Allee, V., Schwabe, O.: Value Networks and the True Nature of Collaboration. Digitial edition edn. ValueNetworks, LLC (2011)
7. Zwickl, P., Reichl, P., Ghezzi, A.: On the Quantification of Value Networks: A Dependency Model for Interconnection Scenarios. In: Cohen, J., Maillé, P., Stiller, B. (eds.) ICQT 2011. LNCS, vol. 6995, pp. 63–74. Springer, Heidelberg (2011)
8. Porter, M.E.: How Competitive Forces Shape Strategy. Harvard Business Review 102 (1979)
9. Von Neumann, J., Morgenstern, O.: Theory of Games and Economic Behavior. Princeton University Press (1944)
10. Breiman, L., Friedman, J.H., Stone, C.A., Olshen, R.A.: Classification and Regression Trees, 1st edn. Chapman and Hall (1984)
11. Von Mises, L.: The Theory of Money and Credit. Jonathan Cape (1934)

Evaluating Impacts of Oversubscription on Future Internet Business Models

Anand Raju[*], Vânia Gonçalves, Sven Lindmark, and Pieter Ballon

IBBT-SMIT, Vrije Universiteit Brussel, Pleinlaan 9,
1050 Brussels, Belgium
{vania.goncalves,pieter.ballon}@vub.ac.be
{Anand.Raju,Sven.Lindmark}@ibbt.be

Abstract. Network oversubscription has long been used by Internet Service Providers (ISPs). While high oversubscription ratios can hamper the end user experience, low oversubscription rates may result in an under-utilization of resources. This paper investigates the impacts of oversubscription, both from a technical (OpEx, Energy footprint etc.) and from a value network point of view (Control and Value creation etc.). A multi-parameter sensitivity analysis of a power model is performed to establish that the choice of oversubscription ratio by an ISP can have a serious impact on their operational energy and capital expenditures. As a next step, a set of business model parameters are operationalized in order to evaluate and establish long-term and short-term impacts of network oversubscription on business stakeholders. Key findings include that there is a need to establish a fit between the technical and business gains of network oversubscription, and that is possible only when an ISP leverages its control and influence over its customer base to better understand and anticipate the network usage, thereby being able to promptly adapting the overall network oversubscription ratios.

Keywords: Future Internet, Business Models, Impact Assessment, Oversubscription, Internet Service Provider.

1 Introduction

Internet Service Providers (ISPs) often oversubscribe their inter-connection network by overbooking the shared infrastructure among their customers. While being cost-efficient, it is also one of the reasons for lower access rates delivered to customers as compared to the rates advertised in their Service Level Agreements (SLAs). As a result, there has been an ongoing, intensely political debate over the nature and impacts of network oversubscription in the literature, much of which considers the choice of oversubscription as a business decision and not a technical decision [5]. However, in reality, oversubscription can be seen as an effective tool used by ISPs to mitigate such implications by finding a fit (in terms of optimal oversubscription ratio) that is symbiotic to end users and long-term economic goals, and at the same time confirming their corporate social responsibility.

---

[*] Corresponding author.

Z. Becvar et al. (Eds.): NETWORKING 2012 Workshops, LNCS 7291, pp. 105–112, 2012.

So far, little work has been done to systematically and objectively map the individual ISP's direct incentives to oversubscribe (e.g. related to energy consumption and operational expenditure) and the impacts of oversubscription on all stakeholders within the ISP's value network. This paper aims to operationalize the impacts of network oversubscription on a single firm, i.e. an ISP, in terms of energy consumption and operational expenditures, as well as, the impact on stakeholders pertaining to the ISP's value network in terms of value creation and control.

In an attempt to provide a holistic view of impacts of choosing and adapting to an optimal oversubscription ratio of a single ISP, Section 2 introduces a comprehensive multi-parameter sensitivity analysis of access power model. It reveals that the choice of oversubscription ratio by an ISP can have a serious impact on the energy footprint of its operation in a given service area. In Section 3 we introduce and elaborate a business model ontology, which is then used for assessing the business impacts of network oversubscription. Building on the results, we then operationalize the business model parameters to evaluate the control and value of an ISP within its value network. The main conclusions are summarized in Section 4.

2   Impacts of Oversubscription on an ISP

Considering the rapid development in design and delivery of access technologies, estimating the power consumption using a generic power model becomes a challenging proposition. These power models [1] are likely to be misinformed as they often fail to assess the impact of underlying variables like oversubscription present in the network. In this work, instead of proposing a power consumption model for an access network, we adapt one of the power models proposed in [1] as represented in (1) and estimate the extent to which the choice of oversubscription by an ISP impacts the overall operational energy footprint of his deployments and resulting savings due to reduction in energy bills and carbon footprint etc.

$$\frac{P_{FTTEx}}{N} = (P_{NT} + 2.\frac{P_{RN}}{N_{RN}} + 2.\frac{P_{CN}}{R_{CN}}) \qquad (1)$$

The evaluation metric (P/N) here expresses the power-per-subscriber as a function of number of subscribers and power consumption of access equipment in a particular region. $(P_{NT}; N_{NT})$, $(P_{RN}; N_{RN})$ and $(P_{CN}; N_{CN})$ represent the power consumption and subscriber base sharing the Network Termination, Remote Node and Central Node respectively. An additional factor of 2 is introduced in order to account for additional cooling and power losses. According to [7], the power model for an FTTEx type of deployment in (1) is reduced to the following:

$$\frac{P_{FTTEx}}{N} = (P_{NT} + 2.\frac{P_{RN}}{N_{RN}} + 2.\frac{P_{CN}}{\frac{a.R_{CN}}{r.N_{RN}+a.r_p}}) \qquad (2)$$

Model also assumes the network aggregation rate (a) as unity and the overall capacity of a Central Node Switch (R_{CN}) is 400Gbps. In order to attain an advertised access rate R (=20Mbps), the throughput capacity of each port is estimated as r_p = 1Gbps. In order to benchmark and cross compare the impacts the power ratings of P_{NT}, P_{RN} and P_{CN} are recorded as 7W, 1465W, and 3000W respectively from Table1 in [7]. However it is to be noted that the power ratings taken directly from the device manual often provide the nameplate rating instead to actual power rating, which may add further sensitivity to our analysis. Finally, the number of subscribers served by each network element is approximated to 960 for a remote node (N_{RN}).

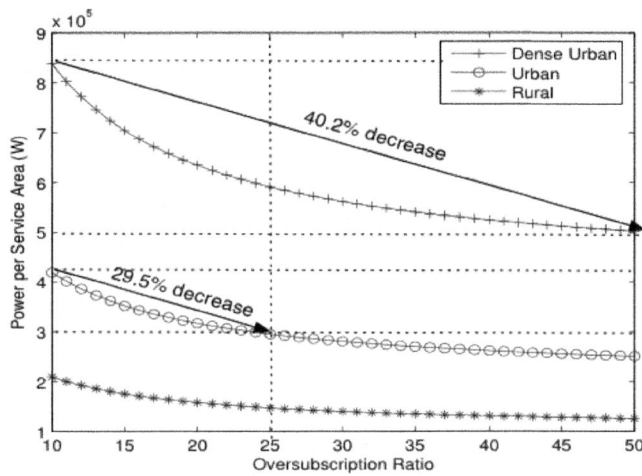

Fig. 1. Power-per-Subscriber in a service area in Watts (y-axis) as a function of variable Oversubscription rates (x-axis)

Oversubscription in access network is introduced when the network configuration meant to serve N subscribers with a peak access rate R, instead supports M ($\geq$ N) subscribers thereby sharing the allocated bandwidth intended for N subscribers [1], resulting in an oversubscription ratio of $\rho = \frac{M-N}{N}$. Assuming R (20Mbps) as the advertised access rate, we have $R = r (\rho + 1)$, where r is the average bit access rate. Using the power ratings, power-per-subscriber metric and oversubscription ratio in the equation (2), we have:

$$\frac{P_{FTTEx}}{N} = \left(25 + \frac{288}{\rho+1}\right) \qquad (3)$$

Consistent with the findings in [1], we delimited the range of the oversubscription ratio from 10 to 50, however it must be noted that some aggressive oversubscription models employed by ISPs may have oversubscription ratios up to 200. Figure 1, gives an overview of sensitivity analysis performed on the power model in response to the parameters ρ and N. The Y-axis represents power-per-subscriber in a service area with subscriber density N varying from ($2^{14}, 2^{13}, 2^{12}$) for a (dense urban, urban, rural) subscriber base. We summarize our findings in Table 1, where an increase in the oversubscription ratio from 10 to 50, reduced the power consumption in a service area by 40 percent. Further increasing the oversubscription ratio to 200, the overall energy footprint reduces to 48 percent of original value.

Table 1. Impact of Oversubscription in terms of subscribers (N= 2^{12})

Oversub. Ratio	Rural (Watts)	Urban (Watts)	Dense Urban (Watts)	Greening (%)
10	51.1(N)	51.1(2N)	51.1(4N)	-
25	36.0(N)	36.0(2N)	36.0(4N)	29.5
50	30.6(N)	30.6(2N)	30.6(4N)	40.2
200	26.4(N)	26.4(2N)	26.4(4N)	48.1

In hindsight, high energy savings may come at the cost of heavy oversubscription and loss of QoS, which may have other business implications (cf. Section 3). Hence, ISPs must be extremely cautious while choosing the oversubscription ratios for their networks. That being said, an intelligent choice of ratio (say ρ = 25) might serve the purpose of providing a network access, which is both energy aware and quality conscious.

3   Impact of Oversubscription on the Value Network Design

While there are many proposed frameworks in the literature for defining individual firms' business models, see e.g. [3][4], the key business model parameters related to control and value creation within the value network are best captured in the business model ontology [2] presented in Figure 2. The framework mainly consists of four abstract layers in which the business models operate under the constraints of three design metrics in each layer. On the one hand, the ontology encapsulates the dimensions of value creation termed as Value Parameters (which relate to aspects such as the value proposition and the financial model). And on the other hand, it captures the functional architecture and value network design parameters termed as Control Parameters.

Control Parameters		Value Parameters	
Value Network Parameters	Functional Arch. Parameters	Financial Model Parameters	Value Config. Parameters
Combination of Assets	Modularity	Cost (Sharing) Model	Positioning
Vertical Integration	Distrib. of Intelligence	Revenue Model	User Involvement
Customer Ownership	Interoperability	Revenue Sharing Model	Intended Value

Fig. 2. Business Model Configuration Parameters

Thus these four layers consist of: (1) The value network: the architecture of actors and roles in the future marketplace; (2) The functional model: the architecture of technical components in the future technological system; (3) The financial model: the

Table 2. Business model parameters and definitions

Business model parameter	Definition
Combination of Assets	Overviews business actors that are in control of resources and assets
Vertical Integration	Represents if the resulting value network would be integrated or disintegrated.
Customer Ownership	Identifies the direct commercial relationship with the customer
Modularity	Refers to the design of systems and artifacts as sets of discrete modules that connect to each other via predetermined interfaces.
Distribution of Intelligence	Refers to the distribution of processing power, control and management of functionality across the system in order to deliver a specific application or service.
Interoperability	Related with the ability of systems to directly exchange information and services with other systems originating from different sources.
Cost Sharing Model	Refers to the anticipated costs for the design, development and exploitation of a product or service
Revenue model	Describes the way a business, monetizes its services and assets
Positioning	Distinguishes between the complementarity and substitutability between products and services
User involvement	Determines the extent to which is customer involvement essential to guarantee adoption
Intended value	Refers to the basic attributes that the product or service possesses which constitute the intended value to be delivered to the customer.

architecture of financial streams determining the future business case; (4) The value proposition: the architecture or general outline of the future product or service. Table 2 elaborates each business parameters captured in Figure 2.

Complementary to the quantitative assessment in Section 2, in the following, we present a qualitative assessment of key business model parameters that are operationalized for long and short-term business implications. In order to examine the implications of oversubscription on the value network design, each business model parameter (see Table 2) is operationalized from a control and value point of view.

Combination of Assets and Vertical Integration: In this context, we consider these main resources as the network elements at the access layer required to provide connectivity to end customers. The ISP would assume full control over these assets and could adjust at all times the maximum number of subscribers served by each network element. Exploiting this control over assets, an ISP can achieve efficiency gains thanks to increase in energy savings, lower equipment deployment and maintenance costs, later translated into operational savings.

Customer Ownership: Traditionally, it is an ISP that interacts and serves the end customers. It also assumes full control over the service and revenue flow from the end customers. In hindsight, not only an ISP has a guaranteed revenue stream but also a clear view on customer behavior and expected demand patterns, using which an ISP can predict the degree of resource utilization and hence can oversubscribe its network more effectively.

Modularity: The network architecture within the access layer is currently to some extent modular and not too complex. The complexity of the network architecture is mainly perceived in the interactions between different layers (access and end user or access and core layers) since management and troubleshooting at one layer is often more difficult when a problem occurs at a lower layer. Greening (lowering the energy consumption in) the network access layer as proposed in Section 2 could lower the resulting QoS. However, problems pertaining to QoS will negatively impact the end customers; its modularity in the architectural design will enable the ISPs to take prompt decisions to assuage these impacts.

Cost (Sharing) Model: Given that present day communication networks are oversubscribed, lower investments in new networking elements can be assumed. Although, our current research does not quantify such savings, we could refer that similar instances of network oversubscription are also prominent in wireless networking cases, where for e.g., in the case of femto cell deployments, impacts of oversubscription on ISP cost structure are clearly identified and estimated in [6].

Table 3. Impact of Oversubscription on ISP Cost Model [6]

Number of users per Femto BS	Area per Femto BS (m^2)	No. of Femto BS (for 10,000 Users)	CapEx (M€)	Savings (%)
4	200	2500	2.50	-
8	400	1250	1.25	50.0
16	800	625	0.62	75.0
32	1600	313	0.31	87.4

With an oversubscribed Femto BS serving up to 32 simultaneous users instead of 4 users, an ISP can save up to 87.4% in terms of capital investments. These substantial gains in CapEx show a high business relevance of network oversubscription at least from an ISP point of view. Still this case differs from the case analyzed in section 2 due to differences in technology, which, for instance, imply that the number of subscribers served by a FTTEx network element (e.g. router) could be higher but at the same time the simultaneity factor would also be higher. This would limit the use of an optimal oversubscription ratio without degrading QoS. Even though we cannot expect the same CapEx savings rate, it is safe to assume that choosing oversubscription in our case would still translate into CapEx savings with a relevant impact on an ISP's cost structure.

Intended Value: The value proposition of greening the network access consists mainly in optimising cost/quality of the service provided. Towards the end customer, the intended value of the service is expressed by a minimum number of outages and delivering the expected quality. Therefore, the key objective lies in choosing the optimal oversubscription ratio, which would guarantee lowering energy costs without degrading QoS. According to the results presented in Table 1, an oversubscription ratio of 25 would satisfy this objective. This value could then be incrementally adjusted according to monitoring activities and customer involvement, resulting in small deviations of power savings depending on subscriber base and period of time. Taking as example the FTTH offer of two French operators (100Mbs advertised download rate), operator A (Figure 3) presents an average oversubscription ratio of 29 while operator B (Figure **4**) presents an average oversubscription ratio of 63. Although operator's B oversubscription is not too high (if compared with ratios of 200), the customers' perceived quality of service could not be acceptable for rich media applications. In addition, the higher the deviation between the advertised and the actual download rates, the higher the customers' perception of unfulfilled QoS.

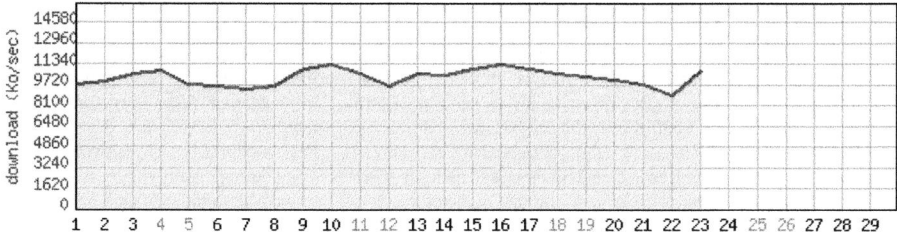

Fig. 3. Monthly sample download access rates of Operator A [8]

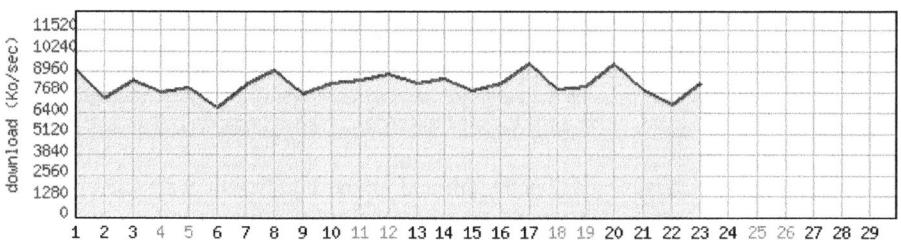

Fig. 4. Monthly sample download access rates of Operator B [8]

4   Synthesis and Implications

The previous sections assessed the impacts of oversubscription both from a technical and a value network point of view. The technical analysis showed that the choice of oversubscription ratio by an ISP has a serious impact on their operational energy and expenditures. In fact, oversubscribing a network can reduce the power consumption for a service area by 40 percent, which in turn has other indirect impacts like reduction of carbon footprint. Therefore, if societal (environmental) costs are also taken into consideration, a higher degree of oversubscription could be preferable than what would have been the case if only based on business decisions.

The business model parameters evaluated in order to establish long-term and short-terms impacts of network oversubscription are summarized in Table 4. One of the key findings of our assessment is that there is a need to establish a strategic fit between the technical and business gains of network oversubscription, this is possible only when an ISP leverages on his control and influence over its customer base to better understand and anticipate the network usage thereby being able to promptly adapting the overall network oversubscription ratios.

In conclusion, given the non-cumulative nature of network bandwidth requirement, oversubscription based business models and design choices will continue to prevail. However, unexpected peaks in demand for bandwidth (due to events like Super Bowl etc.) can critically paralyze the entire delivery mechanism. Also, there is clearly a gap

Table 4. Operationalization of business model parameters

Parameter	Operationalization Strategy	Key Implications
Combination of Assets and vertical integration	Which business actors are in control of networking resources? Would a vertically integrated value network contribute towards positive or negative impacts?	ISP in control over assets allows for achieving efficiency gains. Introducing further actors (e.g. NIP) will likely lead to less oversubscription and lower energy gains.
Customer Ownership	How customer ownership (direct or intermediated) structure looks like? What is the impact of customer ownership in terms of control over the ecosystem?	ISP owns customers, which allows for optimization of oversubscription based on customer behavior and demand patterns.
Modularity	How will the complexity of network architecture and interdependencies between network elements hinder or deter resolving the issues pronounced by network oversubscription?	The modularity of network design (considering mainly access network) allows ISP to balance QoS and savings.
Cost Model	What is the impact in terms of capital investments from ISPs while oversubscribing their networks and serving more customers with limited infrastructure?	Cost savings pertain mainly to the ISPs. In addition to energy cost savings, capital cost will be reduced.
Intended value	From an intended customer value perspective, what are the major impacts for involved stakeholders (ISPs and end users)?	Oversubscription is a key means toward operational excellence, which has to be carefully traded off with QoS.

between headline (advertised) bandwidth and actual bandwidth accessed by the End Customer, and therefore ISPs must take precautionary steps to inform and educate their End Users regarding network oversubscription – its relevance and its importance from technical and business standpoints. Regulators on the other hand, should revise the existing guidelines for marketing the delivery rates by ISPs thereby assuaging and addressing the mutual distrust between the ISP and End Customers.

References

[1] Baliga, J., Ayre, R., Hinton, K., Sorin, W.V., Tucker, R.S.: Energy consumption in optical ip networks. J. Lightwave Technol. 27(13), 2391–2403 (2009)
[2] Ballon, P.: Business modelling revisited: the configuration of control and value. Info. 9(5), 6–19 (2007)
[3] Osterwalder, A.: The business model ontology: a proposition in a design science approach. PhD thesis, HEC Lausanne (2004)
[4] Weill, P., Vitale, M.: Place to Space: Migrating to Ebusiness Models. Harvard Business School Press, Boston (2001)
[5] Sicker, D., Grunwald, D.: Measuring the Network - Service Level Agreements, Service Level Monitoring, Network Architecture and Network Neutrality. International Journal of Communication, 548–566 (2007)
[6] Markendahl, J.: A Tele-Economic Study of Infrastructure Sharing and Mobile Payment Services, PhD Thesis, KTH Sweden (2011)
[7] Gladisch, A., Lange, C., Leppla, R.: Power efficiency of optical versus electronic access networks. In: ECOC 2008, Brussels, Belgium, September 21-25 (2008)
[8] Grenouille, http://www.grenouille.com

Inter-domain Coordination Models

Eleni Agiatzidou[1], Costas Courcoubetis[1], Olivier Dugeon[2],
Finn-Tore Johansen[3], and George D. Stamoulis[1]

[1] Athens University of Economics and Business
{agiatzidou,courcou,gstamoul}@aueb.gr
[2] France Telecom-Orange Labs
olivier.dugeon@orange.com
[3] Telenor Digital Services
finn-tore.johansen@telenor.com

Abstract. In order for the Network Service Providers (NSPs) to provide
end-to-end Quality of Service (QoS) at the inter-domain level different
coordination models have been proposed by ETICS project. In this work
we present and analyse the plausible alternatives of those models and
we compare them with each other in terms of information asymmetry
issues. We show that different information sets affect the total service
provision and we present a basic model analysis on information issues by
means of game-theoretic models.

Keywords: Charging and pricing QoS, game theory, end-to-end
services.

1 Introduction

QoS enabled services such as Gaming as a Service require end-to-end (inter-
domain) connectivity that can be achieved through the existence of agreements
among the involved Network Service Providers (NSPs). Interconnection agreements are constantly changing to fulfil the needs of the new technologies affecting
the Internet architectures [1]. The FP7 ETICS project [2] studies those Internet
architectures and the collaboration models that are capable of sharing information about business and technical parameters to provide QoS-enabled end-to-end
connectivity services.

According to the ETICS solution, the NSPs share information concerning the
traffic source, destination and the statistics of the QoS-enabled service, as well as
the price of their part of the service. The price affects the final service that will be
provided to the buyer. The revelation of the buyer's willingness to pay may lead
to advantageous positions for some providers. On the other hand the total price
that the chain of providers asks for an end-to-end service may lead to failure of
its provision. Also in cases of many paths between a source-destination pair the
pricing strategy of an NSP may affect the final chain of NSPs that will serve
the customer. In this paper, we study a variety of inter-domain coordination
models defined in the framework of ETICS, and we address the information and

Z. Becvar et al. (Eds.): NETWORKING 2012 Workshops, LNCS 7291, pp. 113–120, 2012.

coordination issues that arise. The knowledge of how the information affects the formation of the end-to-end paths is necessary for the right definition of pricing and revenue sharing models.

In section 2 of the paper we describe the ETICS framework while in section 3 we present the different inter-domain coordination models. In section 4, we present a comparison of the models based on information issues. In section 5, we formulate a game-theoretic model pertaining to one of the above models in its simplest form in order to draw some preliminary conclusions that verify our initial analysis.

2   ETICS Framework

The ETICS project [2] has introduced the notion of the ETICS Community. Fig. 1 depicts the interaction between the ETICS community and a customer, who asks for provision of a QoS service from its collaborating NSP inside the ETICS community.

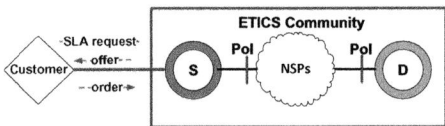

Fig. 1. ETICS Community

The nodes that belong to the ETICS Community collaborate in order to provide QoS-enabled services to members and non-members. Those nodes represent edge, transit or transport NSPs and form paths between an S (source) and a D (destination) NSP. In our considered case, the objective of NSP S (which we will call "buyer" to differentiate from the "customer" which does not belong to the community) is to buy connectivity to an IP range in domain NSP D allowing its packets to reach the specific destination under the conditions desired. Besides the destination range, a Point of Interconnection (PoI) and specific requirements, such as maximum delay, minimum bandwidth, jitter, period of validity etc. have to be defined. The PoI describes where and how the traffic is exchanged between two NSPs. These parameters along with a price specify completely a service that is provided to the buyer, or in other words an SLA offer, which has to be created from sub-SLAs offered by the various NSPs in the path. For the provision of a bundled offer, the NSPs participating in the community have to share information about what they are willing to provide. This is the Publishing Phase. When a buyer is willing to purchase a QoS-enabled service, the Service Composition Phase is triggered to perform the abutment of the different sub-SLA offers in order to provide a global proposal to the buyer. In the next section we study the different scenarios of those two phases and the effect that different information sets may have on the pricing strategies of the participants.

3 Inter-domain Coordination Models

As studied in [3], the role of information asymmetry is critical when considering interconnection agreements. The way information about NSP network capabilities is propagated affects the topology that each NSP will be aware of after the Publishing Phase. The Service Composition Phase can be performed by using Centralized and Distributed models. The Centralized models require the existence of a central entity, called Facilitator, which gathers information from the NSPs and assists the composition of the offer under agreed selection criteria. In the Distributed models the information is propagated between neighbors only. The Service Composition can furthermore be done by using Push and Pull models, which concern the creation time of the offer. In Push models, the offers are available before a buyer's request while in Pull models the offers are created upon a buyer's request. The combination of those models and the creation of two hybrid ones results in six different scenarios. For all these scenarios we investigate the minimum information set that each participant must have available in order for the model to serve its purpose. In all subsequent scenarios, we suppose that an ETICS customer issues requests to the NSP S about connectivity from NSP S to D. Even if we show only one intermediate NSP A for the clarity of the figures, all scenarios could be extended to many intermediate NSPs.

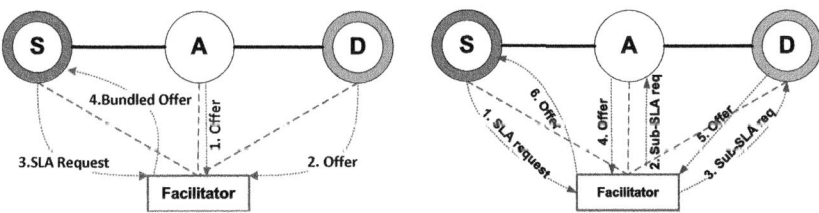

Fig. 2. Distributed Push Model   **Fig. 3.** Distributed Pull Model

Fully Centralized Push Model. During the Publishing Phase all NSPs in the community inform the Facilitator about their sub-SLA offers. For the rest of our analysis we assume that these sub-SLA offers contain: the logical point(s) of interconnection (PoI), the destination(s) network(s) prefix, the QoS characteristics, the expiration time of the offer and a price. The Service Composition Phase begins with the buyer's request. The buyer (S) communicates with the Facilitator and reveals only the PoI (where its traffic is taken care of the ETICS Community), the destination and the QoS characteristics that it wishes to have for the specific end-to-end connectivity (SLA Request). The Facilitator, based on the knowledge that it already has due to the Publishing Phase and the SLA request, combines sub-SLA offers of those NSPs that form a chain from the source NSP to the destination NSP(s). If there are multiple possible offers, the model is flexible regarding the selection criteria of the Facilitator. Indeed, it can choose the best offer from a social welfare or the buyer's point of view or even let the buyer decide. The buyer may accept or not the price of the bundled SLA offer(s) presented by the Facilitator (Fig. 2).

Fully Centralized Pull Model. The Publishing Phase of this model is slightly different from the previous one. The Facilitator is aware of the Network Capabilities (PoIs, and QoS characteristics but not the final price offer) of all NSPs of the community. Based on this information, the Facilitator computes one or multiple NSP chains that could potentially handle the buyer's request for PoI, destination and QoS characteristics. When it receives a request from the buyer it sends specific sub-requests to the NSPs in those chains. After receiving the sub-SLA offers, it combines them. The best offer is chosen under one of the criteria that previously mentioned (Fig. 3).

Distributed Push Model. In this model the information that is available to each NSP depends entirely on the Publishing Phase. Each NSP in this phase publishes its sub-SLA offers to all of its neighbors, to a subset of them or to the whole community. In the first two cases each NSP may combine its own offers with all the offers that it is already informed about and propagate the combined offers to other NSPs in order for the information to be diffused to other participants. Thus, each NSP may be notified about already bundled sub-SLA offers that are available in the community in a cascading way. The buyer will negotiate with its neighbor in order to buy a bundled offer (Fig. 4).

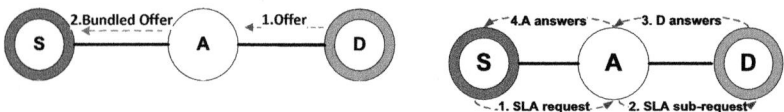

Fig. 4. Distributed Push Model   **Fig. 5.** Distributed Pull Model

Distributed Pull Model. During the Publishing Phase the NSPs exchange information about their Network Capabilities. As in the previous model, the NSPs may decide what kind of information they publish and propagate to each of their neighbors. The Composition phase begins with the buyer's request for connectivity from its neighbor(s) in order to reach a destination (Fig. 5). The neighbor accepts (or rejects) this SLA Request by providing a price offer. In case of accepting, it extracts the part of the requesting SLA that corresponds to its Network Capabilities and adds a price. It subsequently propagates the price and the remaining Network Capabilities requirements to its neighbor(s). After the destination NSP adds a price to the SLA request, the complete offered price is propagated back to the buyer in the reverse cascading way, which then accepts or rejects the final offer. Clearly, the position of an NSP in the chain can affect its strategic power. Such issues can be analyzed by means of game-theoretic models; as discussed in Section 5.

Per-NSP Centralized Push Model. This model is a hybrid of the Centralized and the Distributed Push models. The NSPs create sub-SLA offers (including prices) that are published to catalogues that are available to the community. In the Composition Phase the buyer combines offers and creates a bundled one that fulfils its needs. As opposed to the Distributed model, here the buyer buys the SLA offers from each NSP even if it is not directly connected to it (Fig. 6).

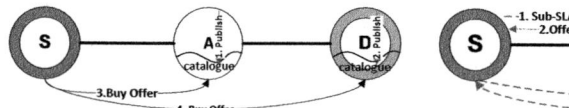 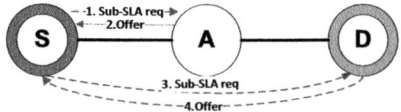

Fig. 6. Per-NSP Centralized Push Model

Fig. 7. Per-NSP Centralized Pull Model

Per-NSP Centralized Pull Model. The buyer is informed about the Network Capabilities that each NSP is willing to provide during the Publishing Phase. During the Composition Phase it asks each NSP in the chain for an SLA offer (which is a part of the whole ASQ good that it wants). In this model (Fig. 7) the buyer communicates separately with each NSP in the chain. The apportioning of price and/or QoS is done according to the buyer's own strategy who leads the composition.

4   Information Issues on the Inter-domain Coordination Models

The total price of the service, the revenue sharing among the NSPs (affected by means of the prices assigned to their sub-SLA offers) and the selection of the NSPs that will participate in providing a service differ across the models. These differences have their origin in the information set that is available to each NSP and is formed during the Publishing Phase, as well as in the fact that a different entity makes the decision of the optimal offer in each model.

In case of a Centralized model the Facilitator has all the available information during the Publishing Phase. In the Distributed models the information must be propagated through the NSPs. The simplest way is to propagate information to the direct neighbors who undertake to bundle it with their own and redistribute it. However, this approach allows for strategic behaviors. For example, an NSP may decide not to propagate the information of all of its neighbors. This may result in different visible options for topologies due to different sets of information available. To overcome this possibility, rules may be enforced in the community in order for the information to be propagated equally. For example a rule could be enforced that the information must be flooded to all community members. Per-NSP Centralized models suffer from the same problem as the Distributed ones. Even if NSPs publish their information in catalogues which can be accessed by every member and thus allowing equal information to all, a minimal set of information about the topology and how the NSPs can be reached has to be available to all participants in the community.

The entity that makes the choice about the optimal offer that will be purchased may be the Facilitator or the buyer. One of the approaches applying to the Centralized models is for the Facilitator to propagate the available choices to the buyer and let it decide; alternatively, the Facilitator may decide according to its own criteria on the final choice, or on a subset of them which are subsequently passed to the buyer for the final selection. In such a case the Facilitator can choose to maximize the expected social welfare, thus improving the overall well-being of the community of customers. Related criteria may be the maximization of the number of buyers that can be satisfied, or of the utilization of the network. Alternatively, the Facilitator can choose the optimal offer by applying potential buyers criteria, typically the lowest price.

In the Distributed (and Per-NSP) models the buyer accepts or not the final offer(s) that are reaching it and thus it has the final choice. However, each NSP may influence the price of the bundled offer. Putting a high price in its part of the service an NSP increases both its profits if this offer is indeed purchased and the risk of rejection of the offer. Authors in [4] present similar problems and show that non-cooperative pricing strategies between providers may lead to unfair distribution of profit. Moreover the position of an NSP in the path to the destination may influence the price that it claims for its part of the service. Thus, certain NSPs may have an advantage in this model due to their position; e.g. the first NSP in the chain, or possibly bottleneck or central NSPs. As this constitutes a problem for all models where the NSPs publish their prices sequentially, and thus most of our models depending on their implementation, we illustrate in the following section the analysis of a simple Distributed Pull model that highlights the effect of the advantageous position of an NSP.

5   Analysis of a Distributed Pull Model

The effect of having an advantageous NSP position in the chain on pricing strategy and on the price of the whole service is presented through a theoretical analysis of a simple Distributed Pull model. We compare the model's efficiency in terms of the probability of the service being offered to the basic collaborative model where the NSPs share the price of the buyer so that they have equal profits.

We assume that we have only three NSPs in the chain; thus, the buyer S requests from A an SLA to connect to D (see Fig. 5). We assume that the Publishing Phase has already been completed and all NSPs have the same set of information about the network capabilities of the other NSPs. The Composition Phase is initiated by S who sends its request for a particular SLA. After deciding on its price P_A, A propagates the SLA and P_A to D. (Note that the results are still applicable if we reverse the roles of A and D, with D choosing first.) D decides its own price P_D and they propagate $P_A + P_D$ to the buyer which takes the offer or not. For A and D to agree to make the transaction their prices have to exceed their costs (C_A, C_D). Also $P_A + P_D \leq P_S$ and $C_A + C_D \leq P_S$, where P_S is the price that the buyer is willing to pay, should apply in order for the service to be ultimately provided. Also, both the costs are here assumed to be known to both the providers who only know the distribution of P_S. The problem that the NSPs A and D have to solve is:

$$maxE\left[(P_*(C_*) - C_*) * 1\ (Service\ is\ provided)\right] \qquad (1)$$

The interaction of A and D corresponds to a Stackelberg game. The optimization problem for D is:

$$max\left[(P_D(C_D) - C_D) * Pr\left[P_A + P_D(C_D; P_A) \leq P_S\right]\right] \tag{2}$$

where P_D is a function of C_D and P_A, s.t. $P_D \geq C_D$. Knowing that D will act this way, the optimization problem for A becomes:

$$max\left[(P_A(C_A; C_D) - C_A) * Pr\left[P_A(C_A; C_D) + P_D(C_D; P_A) \leq P_S\right]\right] \tag{3}$$

where P_A is a function of C_A, and C_D s.t. $P_A \geq C_A$.

In order to proceed with the analysis of a simple yet illustrative case, we assume that P_S follows a uniform distribution and thus $P_S \sim U[0, P_S max]$. Since A chooses its price first it will choose P_A such that $0 \leq P_A + C_D \leq P_S max$. It can avoid the case of $P_A + C_D \geq P_S max$ (where the service is not provided) since it is assumed to know C_D too. (Note that A can be assumed to have some knowledge on the cost of D if e.g. they have similar infrastructure. Of course, the assumption that this knowledge is accurate is only adopted for preserve simplicity of the model.) Once P_A is announced to D, then the only feasible and at the same time meaningful choice for P_D is to satisfy: $P_A + P_D \leq P_S max$. Solving the maximization problem for D we obtain the optimal choice for him (P_D^*). Subsequently, we solve the maximization problem of A. The value of P_D^* is not known to A. It comes as a result of the choice of P_A by A. However, A can make use of the expression of P_D^* and then calculate P_A on this basis. A will choose P_A such that $P_D^* + P_A \leq P_S max$. Through those optimal choices we can calculate the actual profits of A and D in the cases that the service is indeed provided. In the table below, we show some numerical values of those profits and we compare them to the Collaborative scenario. As presented and also proved in closed form the profits of A in the Distributed Pull model are always higher than in the Collaborative one and also double those of D. Also the third row shows a case of a failure of service provision under the Distributed model in contrast to the Collaborative one. The last row shows a failure of both models due to a low price of the buyer.

Also we can evaluate this model by showing the loss in efficiency. This loss can be quantified by comparing: a) the probability $Pr[P_D^* + P_A^* \leq P_S]$ (see eq. 4) that the service is achieved under this model, where A and D follow their own optimal strategies (P_A^*, P_D^*); and b) the probability$Pr[C_D + C_A \leq P_S]$ (see eq. 5) that a service is achieved collaboratively. Thus, when the NSPs act selfishly (Distributed Pull model) there is a huge loss in efficiency, resulting in a reduction of the probability that the service is offered by a factor of 4.

Table 1. Comparison of Profits of A and D

(C_A, C_D, P_S)	Collaborative Model Actual Profits	Distributed Pull Model	
		Actual Profits of A	Actual Profits of D
$(0.043, 0.169, 0.649)$	0.218	0.394	0.197
$(0.113, 0.030, 0.939)$	0.387	0.417	0.208
$(0.075, 0.054, 0.531)$	0.200	-	-
$(0.547, 0.138, 0.149)$	-	-	-

$$Pr[P_D^* + P_A^* \leq P_S] = Pr[\frac{3P_S max + C_A + C_D}{4} \leq P_S] = \frac{P_S max - C_A - C_D}{4P_S max} \tag{4}$$

$$Pr[C_D + C_A \leq P_S] = 1 - Pr[P_S \leq C_D + C_A] = \frac{P_S max - C_A - C_D}{P_S max} \tag{5}$$

5.1   Conclusions

In this paper we have studied the different coordination models proposed by the ETICS project and the way that different information sets affect the service provision under those models. Our research (only partly reported in this paper) has shown that for the Distributed models, either in Push or Pull ones, NSPs may gain an advantageous position against others by excluding a neighbor from having information of other NSPs and thus resulting in a topology different than that in a Centralized model. On the other hand the adoption of a Centralized model implies the concentration of all the information from the NSPs in one single entity which has to be unbiased and meet an optimization goal agreed by all community members. Therefore, future work on the ETICS architecture should consider and evaluate the alternative of using appropriate rules to promote collaboration and deter aggressive selfish pricing strategies.

Acknowledgments. The research work of E. Agiatzidou has been co-financed by the European Union (European Social Fund- ESF) and Greek national funds through the Operational Program Education and Lifelong Learning of the National Strategic Reference Framework (NSRF) Research Funding Program Heracleitus II. Investing in knowledge society through the European Social Fund. Also, this work has been performed in the framework of the EU ICT Project ETICS. The authors would like to thank all ETICS partners for their useful discussions on the subject of the paper and specially M. Dramitinos, N. Le Sauze, H. Pouyllau, P. Zwickl and A. Cimmino.

References

1. Faratin, P., Clark, D., Gilmore, P., Bauer, S., Berger, A., Lehr, W.: Complexity of Internet Interconnections: Technology, Incentives and Implications for Policy. In: 35th Annual Telecommunications Policy Research Conference, George Mason University (2007)
2. 2010 ETICS project consortium, https://www.ict-etics.eu/
3. Constantiou, I.D., Courcoubetis, C.: Information Asymmetry Models in the Internet Connectivity Market. In: 4th Internet Economics Workshop, Berlin (2001)
4. He, L., Walrand, J.: Pricing and revenue sharing strategies for Internet service providers. IEEE JSAC 24(5), 942–951 (2006)

In Which Content to Specialize?
A Game Theoretic Analysis

Eitan Altman

INRIA, Sophia-Antipolis,
2004 Route des Lucioles,
06902 Sophia Antipolis,
France
Eitan.Altman@Inria.fr

Abstract. Content providers (CPs) may be faced with the question of how to choose in what content to specialize. We consider several CPs that are faced with a similar problem and study the impact of their decisions on each other using a game theoretic approach. As the number of content providers in a group specializing in a particular content increases, the revenue per content provider in the group decreases. The function that relates the number of CPs in a group to the revenue of each member may vary from one content to another. We show that the problem of selecting the content type is equivalent to a congestion game. This implies that (i) an equilibrium exists within pure policies, (ii) the game has a potential so that any local optimum of the potential function is an equilibrium of the original problem. The game is thus reduced to an optimization problem. (iii) Sequences of optimal responses of players converge to within finitely many steps to an equilibrium. We finally extend this problem to that of user specific costs in which case a potential need not exist any more. Using results from crowding games, we provide conditions for which sequences of best responses still converge to a pure equilibrium within finitely many steps.

1 Introduction

We consider in this paper competition over content type between Content Providers (CPs). We consider the sitation in which each of several CPs has to decide in what content to specialize. Different types of content may differ by their popularity as well as by the price that idividuals are willing to pay. We may expect that not all CPs will specialize in the most popular content since the income for other content may be larger if its distribution is shared among a small number of CPs.

We first show that this game is equivalent to a congestion game and thus has a potential. Any local maximizer of the potential is thus an equilibrium, and equilibrium can be reached by best response sequences within a finite number of moves [4,2].

Z. Becvar et al. (Eds.): NETWORKING 2012 Workshops, LNCS 7291, pp. 121–125, 2012.
© IFIP International Federation for Information Processing 2012

We extend some of these results to player specific costs in which case there need not be any potential. still a pure equilibrium is seen to exist and we establish convergence to it within finitely many steps under some conditions.

2   The Model

There are M content providers and K content types. We consider below a game in which each content provider has to decide in what content type it specializes. We assume that it cannont specialize in more than one type.

Define a multi-policy $u = (u_1, ..., u_M)$ where u_i is the content type chosen by CP i, taking an integer value between 1 to K.

Let $\gamma^i(u)$ be the utility of content provider i $(i = 1, ..., M)$ when the multi-policy u is used.

Let $\mathbf{m} = \{m_1, ..., m_K\}$ be the vector of the number of CPs specializing in each one of the K content types. We call this the system's state. We shall denote by $\mathbf{m}(u)$ the state that correspond to a multi-policy u.

We shall assume that the utility of CP i for choosing action k depends on the system state $\mathbf{m}$ only through its kth component m_k. i.e. it is a function of only how many CPs (including i) choose content k. We denote this utility as $J^i(k, m_k)$.

Let u be a multi-policy such that $u_i = k$ for some i and k and let $m_k = m_k(u)$. Then

$$J^i(k, m_k(u)) = \gamma^i(u).$$

In the next section we shall assume that $J^i(k, m_k)$ does not depend on i (it ia then omitted from the notation). This is a game with symmetric utilities (or costs). This assumption will be relaxed in the following section.

We consider the set satisfying

$$\mathbf{G}(M) := \{\mathbf{m} : m_i \geq 0, \ i = 1, ..., M, \ m_1 + ... + m_K = M\}$$

Define $S(\mathbf{m}) = \{i : m_i\ 0\}$ to be the support of $\mathbf{m}$, and let e_j denote the unit vector of dimension M with all entries zero except the jth that is one. This set thus contains all possible states in which each of the M CPs has chosen one content type between 1 and K..

Definition. $\mathbf{m}^*$ is said to be an equilibrium in the content game if

$$J(k, m_k^*) \geq J(i, m_i^* + 1), \quad \forall k \in S(\mathbf{m}) \text{ and all } i \neq k.$$

3   Analysis of the Game

Theorem 1. *The following hold:*

- *(i) There exists a pure equilibrium in the content game.*
- *(ii) Define the following potential:*

$$V(\mathbf{m}) = \sum_{k=1}^{K} v(k, m_k), \ \text{where } v(k, j) = \sum_{i=1}^{j} J(k, i)$$

Consider the problem of maximizing the potential V, i.e.

$$Z := \max_{\mathbf{m} \in \mathbf{G}(M)} V(\mathbf{m})$$

Let Q be the subset of $\mathbf{G}(M)$ achieving the max. Then any $\mathbf{m} \in Q$ is an equilibrium in the content game.

Proof. The content game is equivalent to a congestion game as defined in [3], where there is one common source and destination, there are M players and K parallel links and each player has one unit of flow to ship, and has to decide over which link to send it (it is not possible to split the flow between several links). The cost for a player to choose link k if there are m players that choose this link (including itself) is $J(k, m)$. With this equivalence, all statements follow from [3].   ◇

Assume next that the following assmption holds:

A1: $J(k, i)$ is decreasing in i for all k.

Under this assumption the game is not only a congestion game but also a crowding game. As such, we know [2] that that best responses of players converge **within finitely many steps** to an equilibrium (provided that players do not change their strategies simultaneously, and that each player has an opportunity to update its strategy as long as an equilibrium is not reached). This convergence property is called the "Finite Improvement Property" [2].

We next discuss the practical motivaations for Assumption A1 throgh an example. Assume that each CP can satisfy a demand of L downloads per day. Let $D(k, p)$ be the Demand (in downloads per day) for type k content, provided that the price for downloading a unit of such content is p. D is assumed to be strictly decreasing in the price p. Therefore the following inverse function

$$P(k, d) = \{p : D(k, p) = d\}$$

is well defined. It represents the cost of type k content for which the resulting demand is of d downloads per days.

If there are m CPs of type k then the available offer for type k content is mL downloads per day, so that the price of a download that will create a demand that will match this offer is $P(k, mL)$.

The income of a type k CP when there are m CPs specializing in that content is then given by

$$J(k, m) := LP(k, mL).$$

4   Extensions

We study an extension to games with non-symmetric costs and then briefly suggest an extension to the elastic case.

4.1 Player Specific Costs

In [1] we have studied game problems involving two types of content providers: one that corresponds to independent content providers, and one that correspond to content providers that have exclusive agreements with Internet Service Providers (ISPs). The cost for the Internauts who are subscribers of some ISP of fetching content from an independent CP or from ma CP that has an exclusive agreement with another ISP, was assumed to be larger than for fetching it from the a CP that has an exclusive agreement with their own ISP. This impilies that the revenues (and thus the utility) of a CP may depend on its type (independent or not).

This motivates us to allow the utility function that corresponds to choosing some content to be player specific. We thus abandon the assumption of indistinguishability. The game is no more a congestion game and need not have a potential any more. The utility $J^i(k, m_k)$ may depend now on i.

Still the following holds:

Theorem 2. *Assume A1. The following holds in the case of player dependent costs:*

- *(i) There exists a pure equilibrium in the content game.*
- *(ii) In case that there are only two types of contents, the finite improvement property still holds.*

The proof follows Theorems 1 and 2 of [2].

4.2 Other Extension

Both the results of the previous Section as well as the previous subsection have been derived for the case where all CPs participate. We next show how to apply them to the elastic case, i.e. where the offer for content may be a function of the utility that a CP receives. We wish to model a situation in which some CPs may prefer not to participate, if their income goes below some threshold.

We vary the model as follows. We add a new action "0" to each player, which corresponds to having an option of not participating in the game. In addition, we consider a fixed additional cost $c_i \geq 0$ for player i to participate in the game (which may correspond to the investment and maintainance costs). This problem can again be solved with the help of the equivalent congestion game, in which we add an additional parallel link of a fixed cost 0, and where we add c to the cost of each other link. Thus the same results as obtained previously still hold.

5 Conclusions

We have shown how various problems related to competition over content can be reduced to congestion games and/or to crowding games. This allowed us to derive the structure of equilibria and convergence to equilibria within finitely many improvement steps.

Acknowledgement. This work was partly financed by the ARC Meneur of INRIA and by the Alcatel-Lucent INRIA joint Lab.

References

1. Jiménez, T., Hayel, Y., Altman, E.: Competition in Access to Content. In: Bestak, R., Domenech, J. (eds.) NETWORKING 2012, Part II. LNCS, vol. 7290, pp. 211–222. Springer, Heidelberg (2012)
2. Milchtaich, I.: Congestion Games with Player-Specic Payoff Functions. Games and Economic Behavior 13, 111–124 (1996)
3. Rosenthal, R.: A Class of Games Possessing Pure-Strategy Nash Equilibria. International Journal of Game Theory 2, 65–67 (1973)
4. Monderer, D., Shapley, L.S.: Potential games. Games and Econ. Behavior 14, 124–143 (1996)

Estimation of Expectable Network Quality in Wireless Mesh Networks

Till Wollenberg*

Department of Computer Science,**
University of Rostock, Germany
till.wollenberg@uni-rostock.de

Abstract. Our work aims to improve the usability of wireless mesh networks as communication layer of smart office environments. While wireless mesh networks are well-suited for this task in general, the negative impact of interference, fading, and saturation makes the communication basically opportunistic. Our goal is to develop a system which allows a short-term estimation of network quality in terms of throughput, packet loss and latency. The estimation is based on channel measurements and detected high-level activities. With the estimated quality, we expect a significant improvement to user assistance and a speed-up of device integration. In this work-in-progress paper, we define the fundamental problem and its background and design a system that is suitable to solve the problem. We also present some preliminary results from ongoing experiments we carry out in a custom indoor test bed.

Keywords: wireless mesh network, network quality, prediction, smart environments.

1 Introduction and Background

The work presented in this paper is part of a larger research project which focuses on user assistance in *smart environments* (SE). These environments are composed of an ensemble of various devices, many of them mobile or portable. The application scenario covers instrumented rooms—referred to as *smart meeting rooms*—that support teams in tasks such as knowledge exploration and knowledge integration. The present devices include laptops, smart phones, projectors, lights, motorized blinds and canvases, sensors for light intensity and temperature, cameras, microphones and indoor positioning systems.

These devices need to cooperate spontaneously in order to assist the user. The assistance is based on the user's inferred intentions. For achieving their joint goal,

---

* The author is a PhD student funded by the German Research Foundation (DFG) as part of the research training group 1424 "Multimodal Smart Appliances for Mobile Applications" (MuSAMA).

** Postal address: Universität Rostock, Fakultät für Informatik und Elektrotechnik, Institut für Informatik, Lehrstuhl für Informations- und Kommunikationsdienste, D-18055 Rostock.

Z. Becvar et al. (Eds.): NETWORKING 2012 Workshops, LNCS 7291, pp. 126–132, 2012.

it is inevitable that the devices form a network and communicate among each other. According to [7], network concepts suitable for the task must provide a flexible and robust communication layer which (I) does not depend on a central infrastructure, (II) supports device mobility within the network, (III) integrates new devices seamlessly, and (IV) allows the discovery of devices and services.

Wireless mesh networks (WMN) based on IEEE 802.11 Wireless LAN fulfill almost all of the requirements and are therefore well-suited for this task [1]. The technology is commonly available and achieves a throughput which is comparable to Fast Ethernet when using 802.11n. WMN require routing mechanisms that compensate for the inherent properties of wireless communication such as time-varying quality of links as well as movement, arrival, and disappearance of devices. Numerous different approaches exist which implement the mesh functionality either on the network layer (such as OLSR-LQ) or on the link layer (such as 802.11s).

The link quality of the wireless links has a significant effect on the network quality in terms of packet loss, throughput, and latency. Therefore, the applied routing metric must take the current link qualities into account in order to find optimal routes. As to that, numerous cost functions have been developed in the past [2]. The main reasons for the variations in link quality are (a) fading due to attenuation and multipath propagation, (b) interference caused by non-WLAN devices, and (c) transmissions of neighboring WLAN devices.

The 2.4 GHz and 5 GHz frequency bands which are used for WLAN are also utilized by various other wireless systems which normally cannot decode WLAN transmissions (b). Thus, cooperative sharing of the wireless channel does not work here. Additionally, transmissions of other WLAN devices can degrade the link quality significantly (c) because of collisions and the shared medium access.

To enforce certain service qualities in wireless networks, different methods have been proposed. Both or either of prioritization (such as in 802.11e) or reservation (such as in [3]) are commonly used. However, these approaches work neither in the presence of interference caused by non-WLAN devices nor in case of WLAN stations which do not support the corresponding protocols. In the context of our work, we therefore consider the communication in WMN to be basically opportunistic.

2   Goal

Our research project aims at improving the usability of WMN for smart office environments. In these environments, devices execute autonomously planned action sequences. In order to execute a particular action, the network requirements of this action must be met. For the reasons stated in § 1, we can neither use reservation nor prioritization to guarantee this. Our goal therefore is to estimate the short-term network quality. For this, we need to create empirical models that allow us to determine the expectable network quality in terms of throughput, packet loss, and latency.

The estimated values will be used as an input to the strategy synthesis. The main challenge of the strategy synthesis component is to generate device action

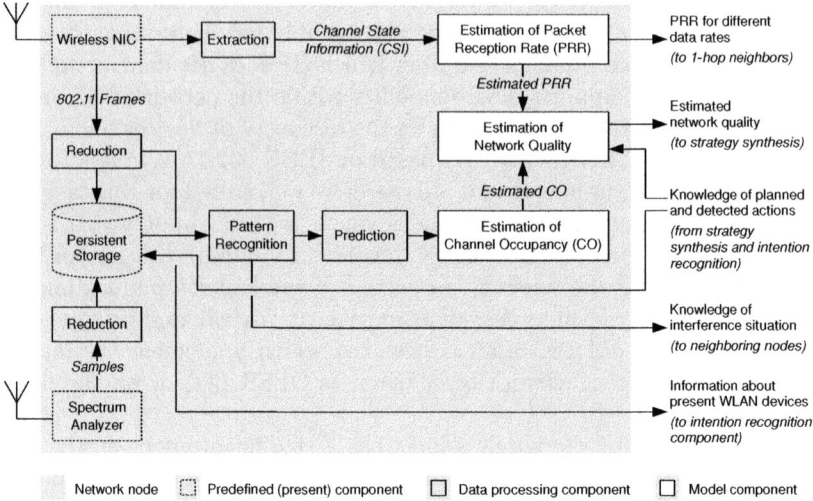

Fig. 1. Overview of the designed system. One node of the mesh network is shown. The labels on the right side indicate information exchange with other nodes of the WMN and other components of the smart environment.

sequences which are then executed by the devices of the ensemble. Generated sequences need to be validated in order to decide whether they are executable. The estimation of network quality is a valuable input in this validation process.

For example, a remote user wants to participate in a meeting which takes place in a smart meeting room. With respect to the user experience, the system should only offer a video chat with the remote user if the network conditions currently allow a stable video transmission. Otherwise, the system may offer only an audio transmission. Also, it is possible to exclude certain actuators during planning if it is known that control commands will not reach these actuators reliably because of high packet loss conditions.

Moreover, the models to be developed will help us to reduce the time needed to integrate a new device into the network. Currently, routing protocols commonly measure the link quality be sending and counting test packets. We can omit this time-consuming process if we can estimate the link quality after only a few exchanged packets. Finally, the experiments we carried out in order to create and validate the empirical models are also helpful to understand the real-world patterns of interference and network usage in smart office environments.

3   System Description

Based on the requirements outlined in §1 and §2 we designed a system which allows us to estimate the quality of a link in the mesh network from (i) knowledge about the intensity and frequency of interference, (ii) the activity of all neighboring WLAN devices, (iii) information about detected user activities, and

(iv) the current channel state. If such information is available for all links within a path between two communicating devices, we can estimate the quality for the entire path. For (i–iii), we consider not only the current conditions, but use also recorded, historic data for prediction. For (iv), we consider only the current state since even minor movement of devices or obstacles can influence the channel significantly due to the strong effects of multipath propagation in indoor environments. A prediction is therefore not promising here.

We designed the system in respect to the possibilities of commercial off-the-shelf hardware to allow straightforward prototypic implementations. Also, we intend the system to work decentralized, as routing in mesh networks is usually organized in a distributed manner. Therefore, one instance of the system is required to run on each network node. Figure 1 shows one instance. Both the wireless network interface card (NIC) and a separate spectrum analyzer serve as data sources. For each received frame, we extract and store the 802.11 frame type, source address and duration[1]. Additionally, we extract the channel state information similar to [5]. To detect interference and non-WLAN spectrum users, we use a simple spectrum analyzer (Ubiquiti AirView). It detects the signal energy across the entire 2.4 GHz ISM[2] band with a resolution bandwidth of 500 kHz. Existing solutions [8] show that this step may be done as well solely with a wireless NIC.

In the recorded data about interference and WLAN activity, we extract patterns and use them for prediction, resulting in a probabilistic estimation of the channel occupancy. In this step, we combine methods of time series analysis with machine learning methods such as Hidden Markov Models. Our goal is to combine information from both observations of spectrum use and knowledge of the user's activities. We classify activities into e.g. "lecture", "group meeting", and "empty room". Eventually, we can describe the channel occupancy rate such as shown in [6], [9] and [4]. We use the current channel state information (CSI) to estimate the probability of a successful frame reception for each 802.11 data rate[3]. Since models for this purpose already exist, this step is not in our focus.

From the estimations of packet loss, channel availability and information about planned or ongoing activities for which the impact on the network is known, we can finally estimate the short-term network quality and supply this information to the strategy synthesis component. Moreover, we are able to provide information about the number and type of currently present wireless devices to the intention recognition component of the SE.

Next to the information exchange between a node and components of the SE there is an exchange among the nodes of the network. First, neighboring nodes can share their knowledge about current and past interference conditions and channel occupancy since we can assume that the conditions are similar for nodes spatially close to each other. Such an exchange is also helpful for new nodes

---

[1] Computed from frame length, data rate and preamble type.

[2] Industrial, Scientific and Medical

[3] e.g. each combination of modulation, forward error correction and antenna configuration

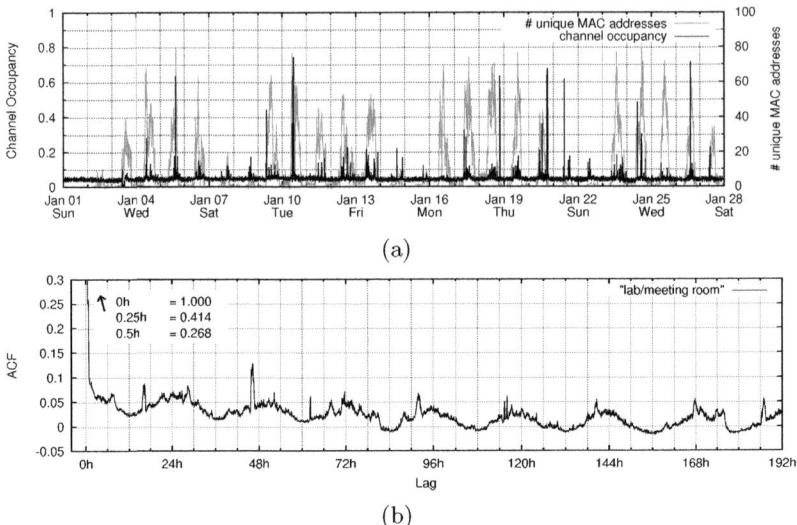

Fig. 2. Time-series data from our indoor test bed. (a) shows probability of occupation for a single channel and the number of active WLAN clients. (b) shows the autocorrelation of the channel occupation from (a). Please refer to § 4 for further details.

joining the network in that they can benefit from the knowledge acquired by nodes which are already on-site. We have to develop a metric which allows us to decide whether knowledge from a neighboring node is applicable to another node. Secondly, each node can provide the results of the packet loss estimations to its direct neighbors in order to allow them to select an optimal 802.11 data rate for their transmissions.

4   Intermediate Results

Based on a commercial router platform, Ubiquiti RouterStation Pro, we built a test bed with ten prototypic network nodes and deployed them in a typical office environment at our university department building. We conducted long-term experiments in this test bed to explore patterns in channel occupancy, network usage, and interference. We also assessed the impact of multipath propagation on current wireless consumer hardware in a typical office environment.

In Fig. 2a we show some measurements from the test bed about channel occupancy of WLAN channels in the 2.4 GHz ISM band. We took data collected from one of our network nodes which is located in a lab room at the department. The room is occasionally used for meetings as well. The black graph shows the channel occupancy of a single WLAN channel on 28 consecutive days in January 2012. We computed the channel occupancy by taking the samples collected with the spectrum analyzer. We observed a span of 20 MHz around 2.412 GHz (channel 1), which is the channel used by most clients in this room. We applied a threshold of $\gamma = -96$ dBm to distinguish between "idle" and "occupied" channel states and finally computed the average occupancy in 60 s windows. For Fig. 2b we took the time series data from 2a and computed the autocorrelation function (ACF) with a lag of up to 192 h (8 days).

It can be seen that the channel occupancy is very low during the night (ca. 5 %, mainly caused by beacon frames) and moderate to high during the working hours. It can also be seen that different days show different patterns of usage. The ACF shows that there is a periodicity of 24 h, but the low values imply that a prediction of consecutive days and even of consecutive weeks is not promising. We found similar results also for other network nodes in our test bed. These nodes are placed in lecture halls, labs, and meeting rooms. We conclude therefore, that prediction of channel occupancy solely with time-series data is not sufficient, because it will not be accurate enough to predict overall network quality. Our approach is therefore to increase the accuracy of prediction by combining time-series data with external information about user activities.

Figure 2a (green graph) shows the number of active WLAN devices nearby our network node, computed by observing WLAN channels 1–13 using a sequential pattern ("channel hopping"). We consider only data frames originating from a wireless station and compute the number of unique MAC addresses occurring within a 10-minute window from that. The high number of active devices observed can be explained by the fact that transmissions from adjacent offices and corridors, and a from small lecture room were also received by our test-bed node. It can be seen in the raw data that a high number of present WLAN devices at a particular point in time indicates a high channel occupancy at this time[4].

5   Conclusion and Outlook

In §1 and §2, we outlined the general problems with the use of wireless mesh networks as communication layer for smart office environments. Instead of reservation or prioritization, our goal was an estimation of network quality. Based on our findings we developed the system model described in §3.

As a next step, we will develop the estimation models for channel occupancy and overall network quality and verify them in our test bed. In the final step, we will deploy the network nodes in the *smart meeting room* and evaluate the overall prediction accuracy in real-world use cases.

References

1. Akyildiz, I.F., Wang, X., Wang, W.: Wireless mesh networks: a survey. Computer Networks 47(4), 445–487 (2005)
2. Baumann, R., Heimlicher, S., Strasser, M., Weibel, A.: A survey on routing metrics. Tech. rep., Comp. Engineering and Networks Laboratory, ETH, Suisse (2007)
3. Carlson, E., Prehofer, C., Bettstetter, C., Karl, H., Wolisz, A.: A distributed end-to-end reservation protocol for IEEE 802.11-based wireless mesh networks. IEEE Journal on Selected Areas in Communications 24(11) (2006)

---

[4] Please note that in Fig. 2a the channel occupancy is only computed for channel 1, while the number of active clients is computed for channels 1–13.

4. Geirhofer, S., Tong, L., Sadler, B.M.: Dynamic spectrum access in wlan channels: empirical model and its stochastic analysis. In: Proc. of the First Int. Workshop on Technology and Policy for Accessing Spectrum, TAPAS 2006. ACM (2006)
5. Halperin, D., Hu, W., Sheth, A., Wetherall, D.: Predictable 802.11 packet delivery from wireless channel measurements. ACM SIGCOMM CCR 40(4) (2010)
6. Oularbi, M.R., Aissa-El-Bey, A., Houcke, S.: Physical layer IEEE 802.11 channel occupancy rate estimation. In: 5th International Symposium on I/V Communications and Mobile Network, ISVC (2010)
7. Poslad, S.: Ubiquitous Computing: Smart Devices, Environments and Interactions. Wiley (2009)
8. Rayanchu, S., Patro, A., Banerjee, S.: Airshark: Detecting Non-WiFi RF Devices using Commodity WiFi Hardware. In: Proc. of the ACM SIGCOMM IMC (2011)
9. Wellens, M., Mähönen, P.: Lessons learned from an extensive spectrum occupancy measurement campaign and a stochastic duty cycle model. Mobile Networks and Applications 15, 461–474 (2010)

Development of Localization Module
for Various Smart Devices Platforms

Ondrej Krejcar

University of Hradec Kralove, FIM, Department of Information Technologies,
Rokitanskeho 62,Hradec Kralove, 500 03, Czech Republic
Ondrej.Krejcar@ASJournal.eu

Abstract. This paper deals with the localization module extension for various platform to support locating in the interior. Any mobile devices (vehicles) can be equipped with a GPS module for location, but in areas with high density in the building interior or city centers a location using GPS is not suitable due to insufficient signal. The solution for these sites can be seen in use of existing WiFi infrastructure. The result of described paper is software module for mobile device. This localization module communicates with the WiFi module for position evaluation of mobile device based on maps and signal strengths of visible WiFi APs.

Keywords: WiFi, localization, multilateration, Gaussian filter, vector map.

1   Introduction

Modern digital world where we live is Smarter and Smarter, whereas information systems (including mobile devices) are more sophisticated, complex and Smarter. Operation systems are nicer, smoother and can predict user needs what means "Smarter" [9].

Current modern mobile phones are recognized as smart devices so they are called "Smart Phones" (Classic design of mobile phones is going to side track). There are several different operation systems existing for Smartphone devices as Windows Mobile, Apple iPhone iOS, Google Android, etc. However each one is trying to implement as much Smart capabilities as possible.

By the term mobile device we understand in this paper not only mobile device known as Smart Phone, but all kind of mobile devices as vehicles, robots, etc. These mobile devices are now used for various tasks from basic types (e.g. car now carry a wide variety of activities, from washing) through the sophisticated ones (e.g. automatic machining of materials) to an intelligent robot machines for separate survey and or collection of environmental samples. The mass deployment is now mainly in areas of human activities that are stereotyped and are relatively easy to replace by automate, or vice versa, in areas that are inaccessible to humans, eventually dangerous or other reason.

In particular, advances in computer technology and machine perception now makes it possible to construct mobile devices that does not operate stereotype on the basis of

Z. Becvar et al. (Eds.): NETWORKING 2012 Workshops, LNCS 7291, pp. 133–138, 2012.
© IFIP International Federation for Information Processing 2012

a predetermined procedure, but on the basis of interaction with the environment through sensors. These devices are then characterized by some degree of "intelligence" and are capable of sensing data based on a certain level of deciding on further action. The basic problem of autonomous mobile devices is their unique location in space. These devices can be equipped with an array of sensors that enable it to perform various kinds of transformations based on sensors measurement (which provide a position in space).

We can use odometry or different types of distance measurement by ultrasonic or laser sensors. These relative localization methods are unfortunately inaccurate (e.g. accumulation error) and often they cannot determine the position (space symmetry, the same corridor in different floors, etc.). As a solution for such cases an absolute positioning methods can be used.

A prototype of mobile exploration vehicle was developed for testing purposes [5]. The vehicle use a number of sensors for orientation in the 3D space. To determine the absolute position in space, vehicle also use a standard localization based on the GPS module to determine the absolute position in space. Location is based on Multilateration. For correct and accurate location in 2D space it is needed to get a line of sight with at least 3 satellites and a sufficient signal. Civilian GPS uses the frequency of 1575.42 MHz (UHF). The use of this technology is unfortunately limited to the above-mentioned line of sight [1]. To locate in these areas is offering the use of WiFi, or some other technology of signal transmitting [7]. The aim of this work is to develop a universal SW localization module which allows localization of various kinds of mobile devices in the interior just using known WiFi network architecture and known WiFi AP position in 3D space.

2   Problem Definition

Wireless localization can be deployed with greater or lesser success in virtually every situation where we have a greater number of broadcast stations, which can be distinguished from each other and determine their distance [6]. Resolution stations can be either based on agent based access with using different transmission channels, carrier frequency, or by a unique identifier (e.g. MAC address) [8].

Location in case of GSM is not sufficiently accurate for its refinement, although it could be used fingerprints, but for this application it is not appropriate.

GPS is already used in the vehicle, but not usable in buildings. Galileo will encounter the same problem, and then again this solution is not applicable.

Localization with RFID requires the installation of a large number of readers, and is useful for monitoring product, but not for "continuous" localization.

Bluetooth and ZigBee achieve similar localization features such as WiFi and are quite commonly used in industry. However, in general the extension is not sufficient and would be pre-installed in the planned deployment destination of the vehicle. It is not acceptable, and for this reason will be selected a WiFi, which can be used today almost ubiquitous infrastructure.

3   New Solution

For the actual location it is especially important to recalculate the value of the RSSI of WiFi signal on the distance from the transmitter [2]. Because the signal is burdened by losses, needs to be filtered before processing. Filtration is carried out with Gaussian filter and calculates the actual position with multilateration [3]. For barrier-free environment, where we expect a smooth sloping curve, can be used to predict losses spread basic empirical model. By modifying the basic empirical model a Lost-Path model can be obtained, which further expands upon the relationship of the antenna gain, loss and slow the spread of losses. Lost-path model is used further as the basis for localization model [4].

4   Implementation

Localization module for the vehicle is the main part of this work. It is written in C++. For debugging under Windows, the application is platform independent, although the target device is based on Linux. The emphasis is on code portability and re-usability. Classes that are not directly related to location core are written in the form of components, independent of other classes and have been tested and tuned separately. The properties of C++ are used and concept of event-driven application is used. In application exist outside the main thread another 4, so taking the effort to stay as long as possible in sleep, that location did not take much of needed computing power for other processes in the vehicle.

4.1   Principles of Operation

The application has two primary inputs and 1 output. The input is the address to the WiFi module and the second input is path to the file with a map. The output is the approximate position with tolerances. When application is started, the map file is loaded into memory and immediately closed. It can be further used by other processes of the vehicle. We created also a parallel thread which periodically reads the information of visible APs and store it to list of pairs MAC address and RSSI values. The list of pairs is forwarded to the process that the steam expands by information from the map. This information is a coordinates, antenna gain, path loss and propagation exponent. If any of this information is missing, the default value is used. You can also enter the following and that least some influence of the environment. Couples who do not have information in the map are removed.

The next stage is to record RSSI history and Gaussian filtering RSSI. The result of this step is a list of AP items, which have position, filtered RSSI value and other necessary data to calculate multilateration. Multilateration first sorts the items according to signal strength and then remove items with a very weak signal. These APs are typically far and the signal is strongly affected by the fraud and bring it calculation error, which is too large. In addition, they can be after the Frances break. It is left to only a certain maximum number of items that the calculation did not last too long. Matrix H and B are created from the remaining items. The calculation is

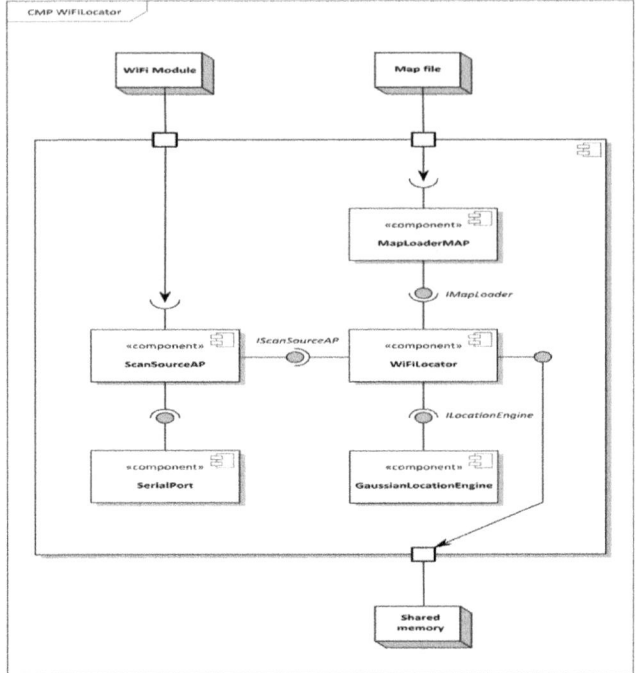

Fig. 1. UML diagram of application components

then performed using the multilateration formula listed above. The application is built on several basic encapsulated classes which are separated by interfaces. For communication between them and between the individual threads are used events. Application architecture demonstrates the diagram of components [Fig. 1].

WiFiLocator - Application is based on the class WiFiLocator. When application is started, instances of classes GaussianLocationEngine, MapLoaderMAP and ScanSourceAP are created. **ScanSourceAP** - The class takes care of communication with the WiFi module. After its launch, the module is switched to AT mode and initializes. **MapLoaderMAP – it** cares about reading and parsing the map. The output is a list (vector) of AP descriptors. **GaussianLocationEngine** - The actual positioning engine. After the engine is called, Gaussian filter is started, and then calculation of locating with multilateration.

5   Testing of Developed Solution

Test of Gaussian filter influence is based on data from measurements with HP iPAQ hx4700. AP was used as the ZyXEL P660HW-T3 device. The [Fig. 2] shows the unfiltered samples and filtered samples with Gaussian filter with different sizes and sigma parameter of kernel.

Simulations showed the assumption that with increasing size and scattering kernel it is a greater suppression of fast losses, but with the growing size of the kernel the system response is worse. The kernel size of 20 is seen that the suppression of leakage

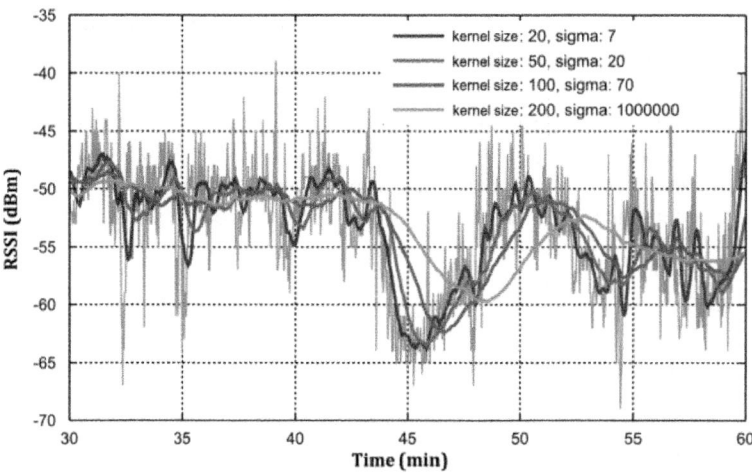

Fig. 2. Application of Gaussian filter to the measured data

is not fast enough, while at 200 the other hand, system response is too slow. As a pretty good compromise it seems to be size of kernel as one hundredth. With this size and zero variance a time delay decrease to half of the size, i.e. 50 samples. This finding implies the need to lay claim to the fastest possible scanning the surroundings and thus the sampling RSSI [10].

5.1   Localization Accuracy

To verify the accuracy, the scheme has been formulated with 4 transmitter (AP). Each transmitter remote from each other about 50 m. The transmitters emit a signal, the attenuation of the load of 10% error with normal distribution on the side of distance. Despite these transmitters were planned route, which passed by a localized point and it did with 10,000 scans. Gaussian filter kernel size is 100 and dispersion 50. The situation demonstrates [Fig. 3]. The black dashed line is the actual trajectory. The blue trajectory is localized and gray points indicate the location, which represent locations without filter.

6   Conclusions

In this paper a development of localization module has been described. This module solves the problem with absolute localization inside buildings. It is a compact and scalable solution with a median accuracy of 3 meters in tested environment. As a source for location data of surrounding APs an extended vector of mobile device map were used. WiFi module which was connected by serial port allows scanning the surrounding visible WiFI AP. The result is the location information including estimated errors. Further direction of this project can be seen in implementation of Monte Carlo localization method, otherwise known as particle filter. This method should itself allow more accurate positioning of WiFi, but also to introduce other information from the sensors to form a single, redundant, robust and above all precise positioning mechanism [10], [11].

Acknowledgement. This work was supported by „SMEW – Smart Environments at Workplaces", Grant Agency of the Czech Republic, GACR P403/10/1310. We also acknowledge support from student Ales Kurecka in development of testing module and in several technical problems they grown during the development phase.

References

1. Brida, P., Machaj, J., Duha, J.: A Novel Optimizing Algorithm for DV based Positioning Methods in ad hoc Networks. Elektronika Ir Elektrotechnika 1(97), 33–38 (2010)
2. Brida, P., Machaj, J., Benikovsky, J., Duha, J.: An Experimental Evaluation of AGA Algorithm for RSS Positioning in GSM Networks. Elektronika Ir Elektrotechnika 8(104), 113–118 (2010)
3. Chilamkurti, N., Zeadally, S., Jamalipour, S., Das, S.K.: Enabling Wireless Technologies for Green Pervasive Computing. EURASIP Journal on Wireless Communications and Networking 2009, Article ID 230912, 2 pages (2009)
4. Chilamkurti, N., Zeadally, S., Mentiplay, F.: Green Networking for Major Components of Information Communication Technology Systems. EURASIP Journal on Wireless Communications and Networking 2009, Article ID 656785, 7 pages (2009)
5. Kotzian, J., Konecny, J., Krejcar, O.: User Perspective Adaptation Enhancement Using Autonomous Mobile Devices. In: Nguyen, N.T., Kim, C.-G., Janiak, A. (eds.) ACIIDS 2011, Part II. LNCS, vol. 6592, pp. 462–471. Springer, Heidelberg (2011)
6. Krejcar, O.: Problem Solving of Low Data Throughput on Mobile Devices by Artefacts Prebuffering. EURASIP Journal on Wireless Communications and Networking – Hindawi 2009, Article ID 802523, 8 pages (2009)
7. Liou, C.Y., Cheng, W.C.: Manifold Construction by Local Neighborhood Preservation. In: Ishikawa, M., Doya, K., Miyamoto, H., Yamakawa, T. (eds.) ICONIP 2007, Part II. LNCS, vol. 4985, pp. 683–692. Springer, Heidelberg (2008)
8. Juszczyszyn, K., Nguyen, N.T., Kolaczek, G., Grzech, A., Pieczynska, A., Katarzyniak, R.: Agent-Based Approach for Distributed Intrusion Detection System Design. In: Alexandrov, V.N., van Albada, G.D., Sloot, P.M.A., Dongarra, J. (eds.) ICCS 2006. LNCS, vol. 3993, pp. 224–231. Springer, Heidelberg (2006)
9. Mikulecky, P.: Remarks on Ubiquitous Intelligent Supportive Spaces. In: 15th American Conference on Applied Mathematics/International Conference on Computational and Information Science, Univ Houston, Houston, TX, pp. 523–528 (2009)
10. Brida, P., Duha, J., Krasnovsky, M.: On the Accuracy of Weighted Proximity Based Localization in Wireless Sensor Networks. In: Simak, B., Bestak, R., Kozowska, E. (eds.) Personal Wireless Communications. IFIP, vol. 245, pp. 423–432. Springer, Boston (2007)
11. Vybiral, D., Augustynek, M., Penhaker, M.: Devices for Position Detection. Journal of Vibroengineering 13(3), 531–535 (2011)

Improved GSM-Based Localization by Incorporating Secondary Network Characteristics*

Marek Dvorsky[1], Libor Michalek[1], Pavel Moravec[2], and Roman Sebesta[1]

[1] Department of Telecommunications, FEECS, VŠB – Technical University of Ostrava,
[2] Department of Computer Science, FEECS, VŠB – Technical University of Ostrava,
17. listopadu 15, 708 33 Ostrava-Poruba, Czech Republic
{marek.dvorsky,libor.michalek,
pavel.moravec,roman.sebesta}@vsb.cz

Abstract. The techniques used in GSM networks for mobile station localization typically use several methods with different level of granularity, based on the base network parameters such as Cell Identification, Timing Advance, the position of Base Transceiver Station, the parameters of Base Transceiver Station antenna etc. This article introduces several others parameters that can be used for network description. This extension can be useful with visualization of localization outputs in cellular network.

Keywords: GSM, localization, TA, RXLEV, SRTM.

1 Introduction

The most basic localisation methods are based on the Cell Identification (CELL ID) [5, 12]. In Global System for Mobile Communications (GSM) networks [1], Base Transceiver Station (BTS) serves only a limited geographical area. The identity of the BTS serving Mobile Stations (MS) provides the simplest location information. An accuracy improvement of this method is achieved if BTS uses sector antennas.

Additionally, the accuracy of MS localization for GSM BTS, which covers areas over a distance of 550 m, can be improved by limiting area by network parameter Timing Advance (TA) [1]. The TA represents the round trip delay between the MS and the Serving BTS (SBTS). According to the GSM specifications [5], the TA is an integer number between 0 and 63. The accuracy of this method depends on the size of the covered area, ranging from hundreds of meters to several kilometers [3]. The main benefit of this method is the independence from the type of MS and a low cost of implementation into the real mobile network [7, 13].

The aim of this paper is to describe an extended network parameters, that can be additionally used in the basic location methods in GSM cellular network. It uses measured values of network parameters that contain information about the current network settings and reception conditions in the SBTS and the six Neighbouring BTS (NBTS) [6].

---

* The work was supported by the IT Cluster 2009-2012 No. 5.1 SPK01/029 grant, funded by the European Fund for Regional Development and Ministry of Industry and Trade, and the paper by the BIOM project, reg. No. CZ.1.07/2.3.00/20.0073, funded by Operational Programme Education for Competitiveness, co-financed by ESF and Czech state budget.

Z. Becvar et al. (Eds.): NETWORKING 2012 Workshops, LNCS 7291, pp. 139–144, 2012.

Recorded information is then sent by each MS during communication over the GSM cellular network.

2 Description of Common Parameters

2.1 Parameters Provided by Network

The present localization methods use a set of common parameters provided by the network. In dedicated mode, the MS periodically sends Measurement Report (MR) messages to the network. These messages contain measurements results of parameters defining reception characteristics from the current SBTS and from NBTS cells. The MS can, at most, report the measurements from the 6 strongest BTS. RXLEV of BTS is measured on the Broadcast Control Channel (BCCH) frequencies and with their Base Station Identity Code (BSIC) a Receiving Quality (RXQUAL) parameters are send through Slow Associated Control Channel (SACCH) to the network. For this purpose, so called measurement report is used, see [4].

The TA parameter is provided by network in periodic intervals. TA is decremented or incremented by an air interface one-bit period of about $3.69\ \mu s$. A one bit adjustments corresponds to a radial distance of approximately $550\ m$).

Furthermore, the MS receives information elements about identity of cell, such as CELL ID, Local Area Code (LAC), Mobile Country Code (MCC) and Mobile Network Code (MNC). The purpose of the Cell Identity information element is to identify a cell within a location area. The purpose of the Location Area Identification information element is to provide an unambiguous identification of location areas within the area covered by the GSM network. [4].

2.2 Parameters Provided by Mobile Telecommunication Operator

Configuration of each BTS is unique and depends mainly on the location of the station. Each mobile telecommunication operator stores a database of records, which consist of following parameters: Longitude, Latitude, Altitude of BTS; CELL ID; BSIC; BCCH; ID of Base Station Controller (BSC); LAC that identify to which location area the BTS belongs; max. electrical power; azimuth of antenna; tilt of the antenna - the angle from horizontal (from -90 degrees (fold up) to $+90$ degrees (fold down)) [1]; the beam width is angular range of the antenna pattern in which at least half of the maximum power is emitted - this angle is then described as beam width or aperture angle (in the GSM network, there are most commonly used antennas with beam width of $60, 90, 120$ and 360 degrees); the antenna Gain.

2.3 Measurement Performed by Drive Test

Drive test is a procedure of recording received levels and timing events whilst driving across a cellular network, e.g. GSM. It is frequently performed during the time when the

---

[1] We distinguish from electrical and mechanical tilt. Electrical tilt is used to reduce signal coverage of a sector area because it improves the close area coverage with increased vertical beam width. BTS mechanical antenna tilt is used as a tool for radio network planners to optimize their networks in sense of coverage and capacity [9].

network is deployed. The reason is that no propagation model can be accurate enough to predict the reception levels in every point of the network. Moreover, the tests are performed also once the deployment of a network is finished. The measurement results can be obtained by performing drive tests with an automatic measurement system, which includes a laptop PC, multi-band GSM test phone and synchronized Global Positioning System (GPS), in order to geo-reference the results.

The trace of drive test should be representative across the network. The trace are chosen according to the concentration of population or according to an importance of place, e.g. motorway. In addition, the measurement results can be also used as an input data for development of a new localization method [8].

3 Abnormal Propagation Areas

We can find areas with abnormal wave propagation conditions in the mobile network. Following text describes a base propagation model used at GSM frequency band, typical problematic situations at abnormal areas with connection to a terrain model that is described by Shuttle Radar Topography Mission (SRTM) data.

3.1 Problematic Places

When determining mobile node location, the method must take into account a wide range of abnormal situations and conditions, where a general approach which uses combination of several methods is likely to fail or may perform worse than a specific method, suitable only for given situation. Some of these most problematic issues are:

- **Signal Reflection** – signal reflections cause problems mainly in mountains and heavy urban areas. The problem is reflected in a wrongly reported TA. Obstacles between MS and BTS cause the MS to receive signal reflected from mountains or on other obstacles. As a result, because of the reflected signal the sector is wrongly visualized with radius that has an equal path length as reflected path.
- **Side Lobes** – one of the antenna parameters is the angle of antenna main lobe α (see chap. 2.2). As a result, the MS is placed aside to main lobe in the direction to so-called side lobe (angle $\alpha + \Delta$). In that case the wrongly visualized sector is limited by α range. In this case it should be detected other antenna sectors and their main lobes.
- **Repeater** – The GSM frequency repeater extends the coverage of a BTS by receiving and re-transmitting signals of the same frequency and at a higher power level. The repeater is a bi-directional amplifier connected to a given BTS.
- **Nanocell** – the Nanocell BTS improves indoor coverage of a cellular system, e.g. GSM. It is cost effective method that improves coverage and capacity in smaller areas such as conference rooms, sheds, halls etc. In addition, nanocells do not require as accurate network planning due to their low power.

We can exemplify the way abovementioned problems influence the localization in an simple example, shown in Figure 1. In this example, we will use only the combination of RXLEV, TA, BTS antenna position and main lobe direction.

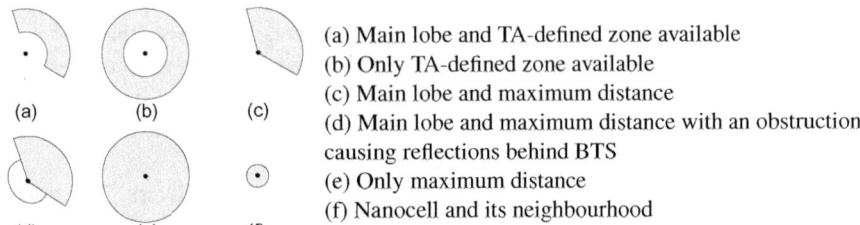

(a) Main lobe and TA-defined zone available
(b) Only TA-defined zone available
(c) Main lobe and maximum distance
(d) Main lobe and maximum distance with an obstruction causing reflections behind BTS
(e) Only maximum distance
(f) Nanocell and its neighbourhood

Fig. 1. Example of CELL ID, antenna main lobe and TA (or RXLEV/max. distance) combinations

When all required information is available, we can use the localization shown in (a) variant. When the antenna main lobe direction is missing, we have to resort to (b) variant. On the other hand, when we do not have the TA value, or expect it to be invalid, we have to resort to variants (c) or (e) if the angle is not available, with RXLEV-based distance limitation, or limiting the distance to technology limits (e.g. 35 km). Should all the parameters be correct, but the reflection is suspected, we can use the pattern (d).

Lastly, should the measured signal come from a nanocell, the distance may be reduced to cover a much smaller area than the TA zone as shown in (f).

3.2   Digital Elevation Models

Digital elevation model (DEM) provides the three-dimensional representation of surface obtained from terrain elevation data. They may be represented as a grid, a set of 3D points or a 3D mesh. In our approach, we would prefer the digital surface model, since it is easier to measure and includes the buildings.

Operators already have to use commercial digital surface models for coverage calculations, in our case we have decided to use one of the freely-available public resources. We chose the Shuttle Radar Topography Mission (SRTM) data[11], which offers terrain elevation data with the resolution of approximately 90 meters (3") for non-US areas. The SRTMv3 provides sufficient coverage for our main goal, which is the calculation of approximate BTS elevation and detection of mountainous terrain.

4   Proposed Secondary Attributes

As mentioned in section 3.1, it is impossible to use a combination of even the basic methods in all situations. Based on available parameters of each individual node in network, we can define a set of parameters which locally describe the whole network. Based on them and expert knowledge, we can define heuristics for correct predicted area visualization.

With the knowledge of network parameters, we can define the first set of secondary attributes:

1. *Average distance of k-nearest NBTSes* – this value can be calculated from network topology, provided by the operator or by obtaining BTS locations form public

resources. We will not include co-located sectors in this calculation and use this value for determination of BTS density.

2. *Intravillan indicator* – based on the Average distance of k-nearest NBTSes, we can determine, if the BTS is likely to be in intravillan or extravillan environment. The threshold between intravillan and extravillan was set to be 3 km, which corresponds to conducted experiments.

3. *Intravillan and Extravillan zones* – calculated from the average distances, these indicators may be used to judge the expected location determination precision (especially when using other, independent methods[10]) and for visual improvements in GIS presentation.

4. *Borders* – state or covered area borders may indicate worse localization performance, since measurement reports may contain unknown combinations of BSICs and BCHs, with both national (not recorded) and foreign BTSes. Also, we may not benefit from techniques based on the inclusion of neighbor lists in this case.

5. *Co-located neighbor indication* – we should identify, if additional sectors of given site are returned in measurement report. In such case, we should also distinguish between sectors co-located in clockwise and counter-clockwise directions. This information can be used for shifting of the predicted angle (or gradient-based visualization), as well as detection of very close proximity to the BTS (based on RXLEVs and other information in measurement report) or excessive reflections, should co-located sectors in both directions be present.

From the BTS position, antenna azimuth and DEM data, we can determine following two attributes:

6. *Antenna elevation* – can be used for both BTS visualization (and better operator decision-making) and signal propagation models (especially if the altitude has not been supplied by the operator).

7. *Terrain granularity* – based on the terrain (and corresponding DEM data) in the direction covered by main antenna lobe, we can determine the terrain granularity. With this information, we can expect and detect the problem. The inspected area can be further fine-tuned by the TA and RXLEV values, similarly to the methods used in methods for modeling of signal propagation [2]. For instance, this can be used to detect mountainous terrain to prevent invalid localizations by using patterns such as Figures 1b and 1d.

Finally, we can include information from actual measurement reports both when doing reference measurements and during the actual localization [10]. Based on these measurements we can define following remaining parameters:

8. *Suspicious measurements* – BTS data measured during data collection phase together with positions may indicate BTSes where invalid TA values or azimuths were measured. This may be used to prevent invalid visualization and we may use this together with a list of repeater CELL IDs.

9. *Broken MS measurement* – some values in measurement report may indicate, that it was broken (e.g. the service BTS appears more than once), mostly during the handover. In this case, we have to decide, which pattern from Figure 1 will be used.

10. *Nanocell indication* – the information may be either provided by operator by special BSC, or by measurements indicating close proximity of BTS but a low RXLEV.

5 Conclusion

In the paper we have developed the set of secondary attributes which will be used together with basic network parameters provided by operators for the selection of correct localization methods and their combination. The resulting attributes could be used as a basis for visualization techniques. In the next step, these attributes will be combined with parameters from other localization method and a resulting model, based both on patterns from Section 3.1, neighboring cell information and vector space model approach [10] will be used to generate the final area.

In future, we could also define additional attributes or use BTS coverage models to better identify possible locations, especially for the UMTS networks, which lack the TA indicator.

References

1. Ahson, A.S., Ilyas, M.: Location-Based Services Handbook: Applications, Technologies, and Security. CRC Press (2010)
2. Backman, W.: Signal level interpolation for coverage area prediction. In: IEEE 60th Vehicular Technology Conference, VTC 2004-Fall, pp. 67–71 (2004)
3. Eberspächer, J., Vögel, H., Bettstetter, C., Hartmann, C.: GSM-Architecture, Protocols and Service, 3rd edn. John Wiley & Sons, West Sussex (2002)
4. Digital cellular telecommunications system (phase 2+), mobile radio interface layer 3 specification
5. GSM 04.31 version 8.18.0 (June 2007)
6. GSM 45.008 version 9.2.0 (April 2010)
7. GSM 08.08 version 8.9.0 (April 2001)
8. Falcone, F., Escauriaza, I.D., Fernandez, A.V., Mau, F.B.: Performance analysis by measurement results in operating 3g network. Wireless Communication Systems, 671–673 (2006)
9. Niemela, J., Lempiainen, J.: Impact of mechanical antenna downtilt on performance of WCDMA cellular network. In: 4th (IEEE 59th) Vehicular Technology Conference, pp. 2091–2095 (2004)
10. Novosad, T., Martinovic, J., Scherer, P., Snasel, V., Sebesta, R., Klement, P.: Mobile phone positioning in GSM networks based on information retrieval methods and data structures. CCIS (PART 2), vol. 189, pp. 349–363 (2011) cited By (since 1996)
11. Rabus, B., Eineder, M., Roth, A., Bamler, R.: The shuttle radar topography mission a new class of digital elevation models acquired by spaceborne radar. ISPRS Journal of Photogrammetry and Remote Sensing 57(4), 241–262 (2003)
12. Samama, N.: Global Positioning: Technologies and Performance. Wiley-Interscience (2008), Wiley Survival Guides in Engineering and Science
13. Steele, R., Lee, C.C., Gould, P.: GSM, CDMA One and 3G Systems, vol. 1. John Wiley & Sons, West Sussex (2001)

Visualization of Complex Networks Dynamics: Case Study

Ivan Zelinka, Donald Davendra, and Lenka Skanderova

Department of Computer Science,
Faculty of Electrical Engineering and Computing Science,
Technical University of Ostrava, Tr. 17. Listopadu 15, Ostrava,
Czech Republic
{ivan.zelinka,donald.davendra,lenka.skanderova.st}@vsb.cz

Abstract. In this article is discussed novel method of the so-called complex networks dynamics and its visualization by means of so called coupled map lattices method. The main aim of this research is to demonstrate possibility to visualize complex network dynamics by means of the same method, that is used spatiotemporal chaos modeling. It is suggested here to use coupled map lattices system to simulate complex network so that each site is equal to one vertex of complex network. Interaction between network vertices is in coupled map lattices equal to the strength of mutual influence between system sites. Also another results from previous experiments, where dynamics of evolutionary algorithms has been converted to complex network and consequently to CML, are mentioned at the end. All results has been properly visualized and explained.

Keywords: complex networks, network dynamics, coupled map lattices, chaos, and visualization, control of complex systems.

1 Introduction

In this article, we try to merge two completely different (at first glance) areas of research: complex networks and chaotic systems visualization.

Large-scale networks, exhibiting complex patterns of interaction amongst vertices exist in both nature and in man-made systems (i.e., communication networks, genetic pathways, ecological or economical networks, social networks, networks of various scientific collaboration, Internet, World Wide Web, power grid etc.). The structure of complex networks thus can be observed in many systems.

The word "complex" networks [1], [2] comes from the fact that they exhibit substantial and non-trivial topological features, with patterns of connection between vertices that are neither purely regular nor purely random. Such features include a heavy tail in the degree distribution, a high clustering coefficient, and hierarchical structure, amongst other features. In the case of directed networks, these features also include reciprocity, triad significance profile and other features.

Z. Becvar et al. (Eds.): NETWORKING 2012 Workshops, LNCS 7291, pp. 145–150, 2012.
© IFIP International Federation for Information Processing 2012

As an example of another exhibition of the existence of complex networks are evolutionary algorithms, whose dynamics can be visualized like complex network structure and dynamics.

The main idea of our research is to show in this article that the dynamics of complex network can be analyzed and visualized like a coupled map lattices (CML), which is usually used for visualization of the chaotic systems. CML system can be understood like a row (1D case) or field of N×M mechanical oscillators, which are mutually joined and influent themselves via nonlinear coupling parameter ε. Their behavior is then visualized as done in Figure 1. Different levels of colors (or levels of gray) represent different "energy" (phase) of each oscillator.

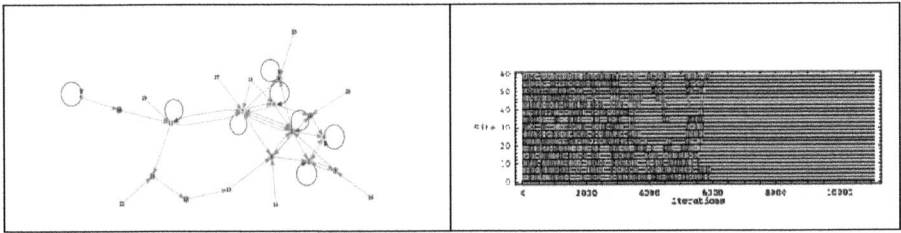

Fig. 1. Example: complex network with more vertices (left), typical CML behavior with chaotic and deterministic "windows", after iteration 6000 it is due to the control fully stabilized (right)

Complex network science is very popular in the last 10 years. Root of it comes from social complex networks, like citation networks etc. During the time it become to be very popular and interdisciplinary important scientific field of research. It has been discovered, that complex network structure and its dynamics, is hidden in the background of many processes that are not, on the first glance, related to the complex networks. Lets note for example classical examples like citation networks, social networks kind of facebook, interactions on the stock exchange, complex communication or power networks, interactions and stress level between people in the closed groups (space flights, submarine missions) etc. are. Another side of coin we can note (from our research) complex network, that is hidden behind dynamics of evolutionary algorithms, see Figure 3 and for full report [3]. This is typical example of formed complex network that show interactions between individuals in the population of the evolutionary algorithm. Each node on Figure 3 is in fact individual from population and links amongst them represent its activities (i.e. which individual has been selected to be parent etc.).

Based on fact, that complex networks and its dynamics can be discovered almost everywhere today, question how to control, analyze or predict them is of course important one. Of course, a lot of standard techniques exist today there. All those techniques, see for example [4], are based on classical mathematical tools, or on heuristic methods, see [5], [6] and [7]. As mentioned before, our aim in this article is to show, that there is quite simple approach how to visualize complex networks and its growth and dynamic in the time so, that well-known methods from chaos control can be used. This approach is discussed in the next section.

2   Complex Networks Dynamics and Its Vizualization

Our method of visualization is based on fact that simplest version of CML (i.e. 1D version) is usually depicted like a row of mutually joined sites, where each site is nonlinearly joined with its nearest sites. Our vision of equivalence between CML and complex network is quite simple. Each vertex is equivalent to the site in the CML. Comparing to the standard CML, sites in complex network CML (CNCML) are **not** joined to the nearest site, but to the sites equal to the complex network vertices. Thus sites in CNCML are not joined symmetrically (i.e. from site X to Y and vice versa) and between different sites is random pattern of connections, which can change in the time. Lets see as an example Figure 1. Vertex No. 12 have feedback to itself (red self-loop), then it influent vertices 7 (double edge), 21 and are influenced by vertices (CNCML sites) 18 and 19. As it is visible, based on our interpretation (CN = CNCML), CNCML is more complex version of the classical simplified version of the CML, however still CML in general, so all techniques of control and analysis shall be working on such a version of CML. Our experiments of CNCML visualization were based on approach that on the beginning has been generated complex network with big number of vertices and no another vertex has been added after all. Only weights on the randomly selected edges have been modified. In all cases has been CNCML calculated and visualized in such a way, that for each CN vertex (i.e. CNCML site) has been calculated according to (5). Different levels of vertices (sites) excitation are depicted by different colors. When compared with Figure 2, it is clearly visible, that our proposed kind of visualization is usable. In Figure 2 and Figure 3 it is observable, that CNCML visualization shows complex and obviously nonlinear behavior of tested CNs. Different kind of behavior (different colors) is clearly visible. Dynamics of the complex network with edges deactivation by real number coefficient. Different colors (gray levels) including white one represent different vertices (sites) activation.

$$S_k = \sum_{i=1}^{N} w_i$$

where   (5)

S_k... is state (excitation or inhibition) of Kth CN vertex (CNCML site)

N... is number of CN vertices

w_i... is weight (random number in general) associated to the each edge.

If L^{th} edge does not exist, then $w_i = 0$

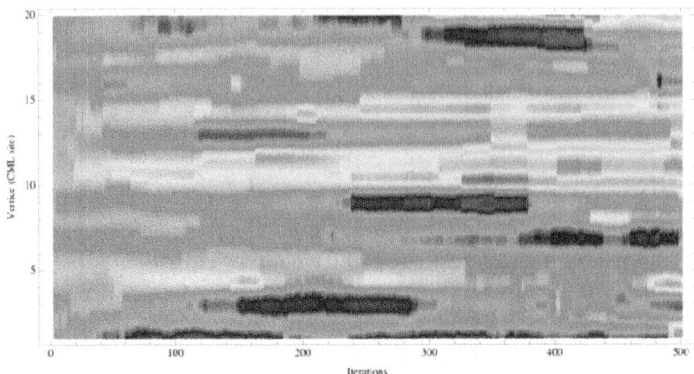

Fig. 2. CML of the network with 20 vertices in 500 iterations

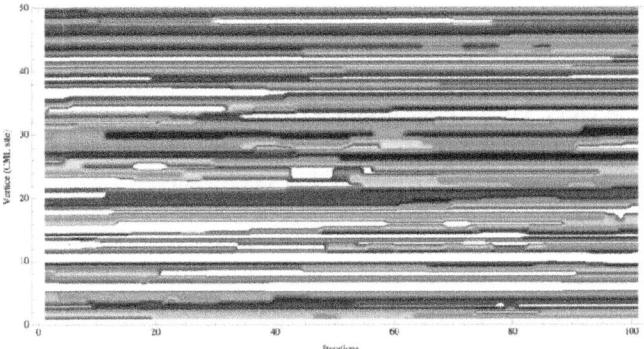

Fig. 3. Dynamics of the complex network visualized like CML system with edges deactivation by real number coefficient. Different colors including white one represent different vertices (sites) activation.

3   Conclusions

In this paper we have proposed method how to visualize and simulate (model) complex network dynamics by means of the CML systems. CML modeling is well known in the domain of so-called spatiotemporal deterministic chaos. For analysis and control of the CML systems has been developed numerous techniques, based on classical mathematics as well as on the heuristic methods. Based on this fact we have used and tested method, how to visualize CN dynamics as a CML system. As a summarization and conclusion can be state this:

- The model of complex network, used for simulations here, has been based on randomly initialized network with different number of vertices.
- Edges between vertices have been also initialized randomly, however, roulette method from genetic algorithms, have been used to add the new edge to the vertex, and/or increase or decrease importance of existing edge. Roulette

method has been used in order to prefer richer vertices (i.e. vertices with more incoming edges) and thus support in this way the main idea of complex networks with small world phenomenon.

- One kind of dynamics for CN visualization has been used. It has been based on preliminary fact, that we already have developed network, with constant number of vertices and dynamic itself depends only on the edge weights evolution.

- All substantial and typical results are visualized via Figure 3 - Figure 2 and it can be stated that our technique of CNCML is usable and complex behavior of CN can be visualized in this way. It is also obvious, that methods of analysis and control of CMLs can be also used for CNCML, because conversion classical CN to the CNCML basically create specific, more complex version, of the CML and there are no restriction of use of contemporary methods on such a version of CML. On the other side, in the case of big computational complexity for complex and big CNCMLs, heuristics, like evolutionary algorithms [8], [5], [6], [7] can be easily used.

Another, more advanced results of this approach are represented in [9]. The main motivation of this research is whether it is possible to visualize and simulate underlying dynamics of an evolutionary process as a complex network. Based on preliminary results (based only on 2 algorithms in 10 versions and 16 test function out of 17) it can be stated that its possible use and benefit of this approach is according to us novel approach to evaluate and control evolutionary dynamics. Based on numerically demonstrated fact (no mathematical proof has been made) that EAs dynamics can be visualized like complex networks we believe that there is new research area for study of EAs dynamics and its possible control via techniques of complex network control [4]. The next research shall be focused on collection of the data from existing real CNs, based on [9] also exact analysis, when chaotic regimes are observable in CNCML, will be done and also what kind of routes to chaos (intermittence, crisis, bifurcations) are there shall be interesting. Another open research issue is to use of selected evolutionary algorithms on CNCML control with "reverse" investigation, what exact impact has such a control on the original CN structure (basically control here means that EAs have to change structure of CN).

Acknowledgement. This work was supported by IT4Innovations Centre of Excellence project, reg. no. CZ.1.05/1.1.00/02.0070 supported by Operational Program *'Research and Development for Innovations'* funded by Structural Funds of the European Union and state budget of the Czech Republic.

References

1. Dorogovtsev, S.N., Mendes, J.F.F.: Evolution of Networks. Adv. Phys. 51, 1079 (2002)
2. Boccaletti, S., et al.: Complex Networks: Structure and Dynamics. Phys. Rep. 424, 175–308 (2006)

3. Zelinka, I., Davendra, D.: Investigation on Relations Between Complex Networks and Evolutionary Algorithm Dynamics. International Journal of Computer Information Systems and Industrial Management Applications (2011) (in print)
4. Meyn, S.: Control Techniques for Complex Networks. Cambridge University Press (2007)
5. Zelinka, I.: Investigation on Realtime Deterministic Chaos Control by Means of Evolutionary Algorithms, In: 1st IFAC Conference on Analysis and Control of Chaotic Systems, Reims, France (2006)
6. Senkerik, R., Zelinka, I., Navratil, E.: Optimization of feedback control of chaos by evolutionary algorithms. In: 1st IFAC Conference on Analysis and Control of Chaotic Systems, Reims, France (2006)
7. Zelinka, I.: Investigation on real-time deterministic chaos control by means of evolution-ary algorithms. In: Proc. First IFAC Conference on Analysis and Control of Chaotic Systems, Reims, France, pp. 211–217 (2006)
8. Zelinka, I.: Real-time deterministic chaos control by means of selected evolutionary algorithms. Engineering Applications of Artificial Intelligence (2008), doi:10.1016/j.engappai.2008.07.008
9. Zelinka, I., Davendra, D., Senkerik, R., Jasek, R.: Does Evolutionary Algorithms Dynamics Create a Complex Networks Structures? Complex Systems 20(2), Champaign, IL 61826 USA

PCI Planning Strategies
for Long Term Evolution Networks

Hakan Kavlak[1] and Hakki Ilk[2]

[1] RAN Network Consulting,
Ericsson Japan K.K, Tokyo, Japan
hakan.kavlak@ericsson.com
[2] Ankara University, Faculty of Engineering, Electronics Engineering Department,
Tandogan, Ankara, Turkey
ilk@ankara.edu.tr

Abstract. In Long Term Evolution (LTE) networks, physical cell identity allocation (PCI) is crucial for quality of service and somewhat similar to scrambling code allocation in WCDMA. PCI, or Layer 1 identity, is an essential configuration parameter of a radio cell. It identifies the cell in mobility functions such as cell reselection and handover. In this paper simulation results, in order to optimize PCI planning in LTE networks is developed and several recommendations for PCI planning strategies are presented.

Keywords: Network planning, Optimization, LTE, PCI Planning.

1 Introduction

As smartphone penetration increase all over the world, mobile users consume more data and wish for faster networks. From speech only GSM to LTE, this demand has to be satisfied. Today, an ordinary LTE network can reach up to 100 Mbit/s. In order to reach this kind of speed, LTE networks should be well optimized. As an initial step; optimization and design part of LTE defines the name of the cells so that users can camp on the best server cell and handover the other cells while the user is moving [3]. One of the basic functions in any network is the cell search. During this procedure, time and frequency synchronization are established between the user equipment (UE) and the network. To identify the cells, the Physical layer Cell Identity (PCI) is acquired. This is achieved by the cell search procedure [1], [8]. The Physical Cell Identity, or Layer 1 identity, is an essential configuration parameter of a radio cell. It identifies the cell in mobility functions such as cell reselection and handover. The PCI is also used to determine the location of the resource elements containing the Physical Control Format Indicator Channel (PCFICH) and Physical HARQ (Hybrid automatic repeat request) Indicator Channel (PHICH) [8]. If cells` PCI cannot be assigned well, mobile users cannot read the actual signals, cannot camp on the LTE networks or data throughput is degraded even worse, it is dropped. That means high cost network becomes useless [13]. PCI consists of two signals; Primary Synchronization Signal

Z. Becvar et al. (Eds.): NETWORKING 2012 Workshops, LNCS 7291, pp. 151–156, 2012.

(PSS) and the Secondary Synchronization Signal (SSS). The detection of these two signals not only enables time and frequency synchronization, but also provides the UE with the physical layer identity of the cell and the cyclic prefix (CP) length, and informs to the UE [2]. For Cell Search procedure UE first detect PSS and then SSS as shown in Fig 1.

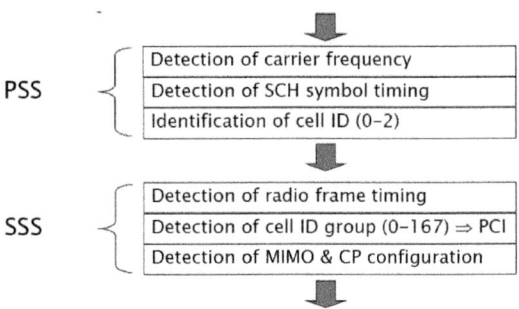

Fig. 1. Cell search precedure

The primary sequence is referred to as the Primary Synchronization Signal. Three PSS sequences are used in LTE, corresponding to the three physical layer identities within each group of cells to estimate 5msec timing and physical-layer identity. It is used to determine S-SCH (synchronization channel) symbol timing. [11] 168 pseudo random sequences represent the Secondary Synchronization Signals. They determine frame timing and the cell identity group. Each SSS sequence is constructed by interleaving, in the frequency-domain, two length-31 BPSK-modulated secondary synchronization codes, denoted here m_0 and m_1. Frame structure of PSS and SSS can be shown in Fig 2.

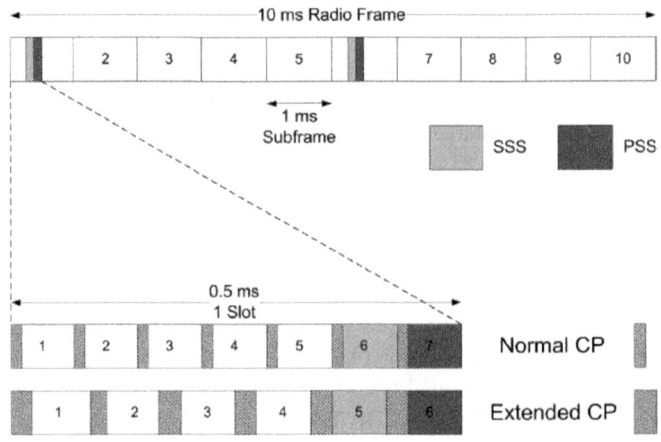

Fig. 2. PSS and SSS frame and slot structure in time domain

PCI is calculated by using PSS and SSS in a formula (1). PCI N_{ID}^{CELL} is defined as

$$N_{ID}^{CELL} = 3N_{ID}^{(1)} + N_{ID}^{(2)} \tag{1}$$

SSS $N_{ID}^{(1)}$ is the physical layer cell identity group (0 to 167) and PSS $N_{ID}^{(2)}$ is the identity within the group (0 to 2). This arrangement creates 504 (168x3) unique physical cell identities. Next section describes PCI structure and creation of PCI signals.

2   PSS and SSS Signal Structure

2.1   PSS Signal

PSS signal consist of 3 different sequences called Physical-Layer Identities. Range is 0 to 2 and 62 symbols long. The sequence $d_u(n)$ used for the PSS is generated according to:

$$d_u(n) = e^{-j\frac{\pi u n(n+1)}{63}} \quad \text{for } n=0,1,\dots,30 \tag{2}$$

$$d_u(n) = e^{-j\frac{\pi u(n+1)(n+2)}{63}} \quad \text{for } n=31,32,\dots,61 \tag{3}$$

where u is the Zadoff-Chu root sequence index and depends on the cell identity within the group $N_{ID}^{(2)}$:

Table 1. PSS sequences

$N_{ID}^{(2)}$	Root index u
0	25
1	29
2	34

2.2   SSS Signal

It is based on maximum length sequences (m-sequences). An m-sequence is a pseudo-random binary sequence which can be created by cycling through every possible state of a shift register of length n, resulting in a sequence of length 2^n-1. Three length-31 m-sequences are used to generate the synchronization signal denoted $\tilde{s}$, $\tilde{c}$ and $\tilde{z}$, respectively. Two length-31 binary sequences are used to generate the SSS. Sequences $s_0^{(m_0)}$ and $s_1^{(m_0)}$ are different cyclic shifts of an m-sequence $\tilde{s}$. The indices m_0 and m_1 are derived from the cell-identity group $N_{ID}^{(2)}$ and determine the cyclic shift.

2.3   Creating PCI

For each cell, $PCI_i = 3SSS_j + PSS_k$

  ∘ $i = 0 \dots 503$
  ∘ $j = 0 \dots 167$ group
  ∘ $k = 0 \dots 2$ ID

The sequence for the SSS signal is generated as follows:

$$\circ\ q' = INT(Sj/30) \tag{4}$$

$$\circ\ q = INT((Sj+q'(q'+1)/2)/30) \tag{5}$$

$$\circ\ m' = S j +q(q+1)/2 \tag{6}$$

$$\circ\ m_0 = m' \bmod 31 \tag{7}$$

$$\circ\ m_1 = [m0+INT(m'/31)+1] \bmod 31 \tag{8}$$

Simulations hint that the following combinations at adjacent cells will give bad performance, i.e. long synchronization times and high interference. But this is still not proven by the field tests.

  ∘ Same ID, i.e. same k
  ∘ Same m_0
  ∘ Same m_1

This is valid for both synchronous and asynchronous networks. Simulations for synchronous networks, shows the difference of times for camping on the network with respect to the signal to interference ratio [6].

3   Simulation Results

To show the synchronous time with respect to changing interference and PCI values a trial lab test was conducted. Details can be found in reference [4]. Simulation assumptions are given in reference [4], [5] and [6]. For different fading scenarios (ETU5, ETU300 and EPA5) cell identification delay computed as the time required by the UE to properly detect the cell by detecting correctly detecting its PSS, SSS and radio framing boundary [7] is illustrated in figure 3 for ETU5 where the user is mostly in this class.

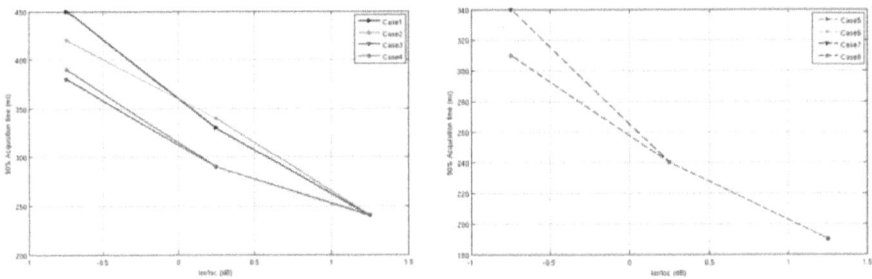

Fig. 3. Cell identification time (ETU5 channel)

These results show that cell detection time is longer in synchronous scenario than in asynchronous scenarios for the same SNR [10]. These results show that PCI planning should have a list of approaches that needs to be defined. The following section give several recommendations.

4   PCI Planning Recommendations

In order to achieve the optimum quality, PCI assignment must be cell identification delay computed as the time required by the UE to properly detect the cell by correctly detecting its PSS, SSS and radio framing boundary Major recommendations about PCI planning try to prevent cells with the same Physical Cell ID from overlapping, take into consideration the neighbor cell relationships in the assignment of Physical Cell IDs, provide a method for reserving codes for use with new LTE infill sites, indoor systems, future LTE Femtocell rollout in the cell identification delay computed as the time required by the UE to properly detect the cell by detecting correctly detecting its PSS, SSS and radio framing boundary codes to be spread over distance and force a uniform use of the codes in the network [9], [4]. With respect to the above recommendations, there are two main strategy options:

1. Random planning i.e. PCI plan that does not consider PCI grouping and does not follow any specific reuse pattern
2. Ordered planning: Neighboring sites are grouped into clusters, and each cluster is assigned a limited number of Code Groups. Each site is assigned a specific Code Group and each sector a specific Color Group

Second one obviously delivers better network performance due to less interference but this method also divided into 2 subsections. One method is giving same SSS and different PSS to the cells in a site. Using the same SSS ID per site allows better management and trouble shooting in the network. This is applicable when we have a maximum of three sectors per site. It also allows optimal synchronization times in the system. This method is also advised by Ericsson because of easy use and allows optimal synchronization times [10]. Second method is more complicated. Use different PSS and also SSS. SSS values can be select with respect to different m_0 and m_1 values.[1] All together it will eliminate the risk of having the same k or frequency shift in the same site, in adjacent cells or pointing at each other. Clustering is another aspect to plan. Clusters can be selected equal to one Tracking Area (TA). Limitation is one TA can't exceed 28 sites. This method can be implemented for using 5 MHz or 10 MHz bandwidth network, ie: For 5-10 MHz network paging load per RBS is limited and small TA is needed. Another method can be creating cluster in a TA, but then other TA borders should be carefully planned. Basic clustering and coloring showed in reference [8].

For neighboring cells PSS values are selected other than the source cell. By using WCDMA WNCS output, neighbor relation can be approximated. At the end if a cell has PSS 0, adjacent cells have PSS 1 and 2 and also best neighbor in other site have PSS 1 or 2 [9].

5   Conclusion

LTE system is novel for many networks. PCI planning is very important and any wrong assignment may cause bad experience to the user. PCI planning looks like WCDMA Scrambling Code (SC) planning. Many simulation results and papers written on it still need field experience. OFDM performs better than WCDMA by shifting some important signaling data in wide frequency band. Because of this change, PCI planning is also more complicated than SC planning [5]. When planning PCI the following priority orders are recommended:

1. The same PCI should be avoided within the same site and as neighbors.
2. PCI with conflicting k values should be avoided within the co-site and neighbors.
3. PCI with conflicting m0 and m1 values should be avoided within the same site and as neighbors

There is no field trial about PCI plan strategy; however the proposed way is the most optimum method to reduce interference and improve the synchronization time. If new tests show there is no or little improvement between the ordered and random method, operator can change it in more expanded network.

References

1. Holma, H., et al. (eds.): LTE for UMTS. Wiley (2009)
2. Sesia, S., et al. (eds.): LTE, The UMTS Long Term Evolution: From Theory to Practice. Wiley (2009)
3. LTE Protocols and Procedures, Ericsson (2009)
4. Performance of LTE cell identification in multi-cell environment, 3GPP TSG RAN WG4, R4-080691 (2008)
5. R4-072215, "Simulations assumptions for Intra-frequency cell identification", Texas Instru-ments, NXP, Motorola, Ericsson, Nokia
6. R4-080479, "Summary of RRM ad-hoc", Nokia Siemens Networks, Motorola
7. Ghosh, A., Zhang, J.: Fundamentals of LTE. Prentice Hall (2010)
8. LTE 10A Air Interface LZU 108 7260 R1A, Ericsson (2009)
9. Kreher, R., Gaenger, K.: LT E Signaling, Troubles hooting and Optimization. John Wiley & Sons (2011)
10. Dahlman, E., Parkvall, S.: 4G LTE/LTE-Advanced for Mobile Broadband. Elsevier (2011)
11. 3GPP Mobile Broadband Innovation Release 9, Release 10 and beyod:HSPA+, LTE/SAE and LTE Advanced (2010)

Effective Packet Loss Estimation on VoIP Jitter Buffer

Miroslav Voznak[1], Adrian Kovac[2], and Michal Halas[2]

[1] VSB – Technical University of Ostrava,
17. listopadu 15, 708 33 Ostrava-Poruba, Czech Republic
miroslav.voznak@vsb.cz
[2] STU – Slovak University of Technology, Faculty of Electrical Engineering,
Ilkovicova 3, 812 19 Bratislava, Slovak Republic
{kovaca,halas}@ktl.elf.stuba.sk

Abstract. The paper deals with an influence of network jitter on effective packet loss in dejitter buffer. We analyze behavior of jitter buffers with and without packet reordering capability and quantify the additional packet loss caused by packets dropped in buffer on top of the measured network packet loss. We propose substitution of packet loss parameter P_{pl} in ITU-T E-Model by effective packet loss P_{plef} incorporating network jitter, a jitter buffer size and a packet size as additional input parameters for E-Model.

Keywords: Jitter buffer, Dejitter buffer, MOS, E-Model, Packet loss, Packet drop; Pareto distribution, Pareto/D/1/K, VoIP.

1   Introduction

Voice over IP communication gains still greater importance in telecommunications industry. Lack of synchronization in comparison to TDM (Time-division Multiplexing) brings concerns about variable conditions on network which cause packet loss and fluctuations in delay (jitter). Jitter causes excess packet loss on receiving buffers depending on the buffer size and delay variance. When using E-Model as an objective method based purely on network parameters for speech quality assessment, effects of jitter are not incorporated in original E-Model. Estimated MOS (Mean Opinion Score) can be positively biased or be too optimistic about call quality under real IP network conditions. In our paper we propose methods of numerical approximation of general jitter buffer behavior. These approximations are used to quantify effective network packet loss P_{plef} taking three additional network parameters on the input: jitter buffer size [ms]; voice packet size [ms] and network delay variance (jitter) [ms].

2   Selected Time-Related IP Network Parameters

Under time-related network parameters we understand packet transmission delay, mean interarrival time difference – jitter and secondarily packet loss, which can be understood as infinite delay of packet delivery. Based on practical experience of voice

Z. Becvar et al. (Eds.): NETWORKING 2012 Workshops, LNCS 7291, pp. 157–162, 2012.

perception and human conversational model, ITU-T G.114 [1] defines the recommended value of Mouth-to-Ear delay which consists of partial delays occurring at different stages of communication path. The delay distribution on IP networks can be successfully modeled and described by Pareto distribution which give better results compared to Weibull, Poisson or Log-Normal distribution [2]-[4], [11]. Pareto distribution belongs to family of geometrical and long-tailed distributions that characterize statistical set where most values are clumped around the beginning of the interval. Equation (1) shows PDF and equation (2) CDF functions of generalized Pareto distribution (GPD).

$$F_{(\xi,\mu,\sigma)}(x) = 1 - \left(1 + \frac{\xi(x-\mu)}{\sigma}\right)^{\left(-\frac{1}{\xi}\right)} \qquad f_{(\xi,\mu,\sigma)}(x) = \frac{1}{\sigma}\left(1 + \frac{\xi(x-\mu)}{\sigma}\right)^{\left(-\frac{1}{\xi}-1\right)} \qquad (1), (2)$$

Where σ = scale, ξ = shape and μ = location parameter (min. value of random variable with Pareto distribution), μ is an offset of Pareto curve from zero and represents minimal network delay $T_{a\text{-}min}$. Jitter J [ms] is calculated in real-time as floating average of 16 samples of differences between interarrival times (timestamps) of consecutively received packets contained in RTP (Real Time Protocol) header, the calculation is defined in RFC 1889 and given by equation (3) where each difference is calculated according to equation (4). Jitter value is transferred in RTCP (Real Time Control Protocol) protocol header as one of the QoS parameters.

$$J = J + (|D_{(i-1,i)}| - J) / 16 \ [ms] \quad \text{and} \quad D_{(i,j)} = (R_j - R_i) - (S_j - S_i) \ [ms] \qquad (3), (4)$$

Where R are timestamps of packet reception time, S when packet was sent and indices i,j are consecutive packet numbers. According to G.1020 [5] an alternative approach of jitter based on determination of the Mean Absolute Packet Delay Variation with regard to a short term average or minimum value (adjusted absolute packet delay variation) offers more accurate calculation short-term jitter better describing relationship to jitter buffer behavior. In [5] the short term jitter is computed for current packet (i) whose delay is designated t_i. Packet (i) is compared to a running average estimate of the mean delay using 16 previous packet delays and assigned either a positive or negative deviation value. These characteristics are calculated according to equations (5), (6) and (7) [5], where a Mean Delay is expressed in relation (5), a Positive dealy deviation in (6) and Negative deviation in (7).

$$D_i = (15 \times D_{i-1} + t_i\text{-}1) / 16 \tag{5}$$

$$P_i = t_i - D_i \quad \text{if } t_i > Di \quad \text{and} \quad N_i = D_i - t_i \text{ if } t_i < Di \tag{6), (7}$$

If $t_i = D_i$, then both P_i and N_i are zero. Mean Absolute Packet Delay Variation 2 ($MAPDV2$) for packet i is computed as (8) [5]:

$$MAPDV2 = mean(P_i) + mean(N_i) \tag{8}$$

Variance of input stream packet delivery causes problems with synchronous playback. Receiver can wait finite amount of time for data to be delivered [12], [13]. All data that are not lost in the network but arrive later than expected are considered lost, hence we use the term *Effective packet loss* to describe all losses in data path. Effective packet loss is encountered at the input of audio decoder and is equal to or greater than measured network packet loss advertised through RTCP packets. We analyze additional loss on jitter buffer and its inclusion into E-Model further.

3 E-Model as Objective Measurement Method

E-Model defined by ITU-T G.107 is widely accepted objective method used for estimation of VoIP call quality [6], [7]. E-Model uses set of selected input parameters to calculate intermediate variable, *R-factor*, which is mostly converted to MOS-CQ value. Input parameters contribute to the final estimation of quality in additive manner as expressed in (9).

$$R = R_o - I_s - I_d - I_{e\text{-}eff} + A . \tag{9}$$

Where R_o represents the basic SNR, circuit and room noise; I_s represents all impairments related to voice recording such as quantization distortion, low voice volume, compression artifacts; I_d covers degradations caused by audio signal delay including side tone echo; $I_{e\text{-}eff}$ impairment factor represents all degradations caused by network transmission path, including end-to-end delay, packet loss, codec compression artifacts and PLC (Packet Loss Concealment) masking capabilities and A is an advantage factor of particular communication technology. We focus at $I_{e\text{-}eff}$ parameter, which is calculated as in (10):

$$I_{e\text{-}eff} = I_e + (95 - I_e) \times P_{pl} / (P_{pl} + B_{pl}) . \tag{10}$$

Where I_e represents impairment factor given by codec compression and voice reproduction capabilities, B_{pl} is codec robustness describing immunity of particular codec against random losses and quantifying its PLC masking qualities. These values are listed for narrowband codecs in ITU-T G.113 appendix [8]. We propose to substitute P_{pl} parameter for an overall effective packet loss P_{plef} according to equations (11) and (12). In this document we refer to "Modified E-Model" as to model based on original ITU-T G.107 with P_{pl} packet loss substituted by proposed P_{plef}. Both P_{pl} and P_{plef} values lies in interval <0,1>.

$$P_{plef} = 1 - (1 - P_{pl}).(1 - P_{jitter}) \quad \text{and} \quad P_{jitter} = (P_{plef} - P_{pl})/(1 - P_{pl}) \tag{11), (12}$$

Where P_{jitter} is net packet loss on jitter buffer. According to the type of buffer used, P_{plef} should be substituted by either P_{loss_wo} or P_{loss_wr} respectively.

3.1 Jitter Effects Simulation and Measurements of Effective Packet Loss

Measurements, which were carried out, proved the jitter buffer behavior described in chapter 2 and helped to find a best fitting function modeling the jitter buffer packet loss P_{jitter}. We have simulated MOS dependence of several codecs such as G.711 both μ-law and A-law with PLC, G723.1 (ACELP and MP-MLQ), G.726, G.729 on QoS parameters which were presented by One-way delay $T_a \in \{0, 20, 50, 100, 150, 200, 300, 400\}$ [ms]; Network packet loss $P_{pl} \in \{0, 1, 2, 3, 5, 7, 10, 15, 20\}$ [%] and Network jitter of 20, 40 and 80 ms with Pareto distribution. VoIP traffic was emulated using IxChariot software and the transport network presented a computer with two network cards with WANem software implementing Pareto distribution on an output packet stream. The stream was received with third computer with IxChariot endpoint, the situation is depicted in Fig. 1. For each codec and combination of QoS parameters values were simulated one-minute VoIP calls and repeated three times, from which data for evaluation were collected, overall 189 combinations of parameters were simulated [9], [10].

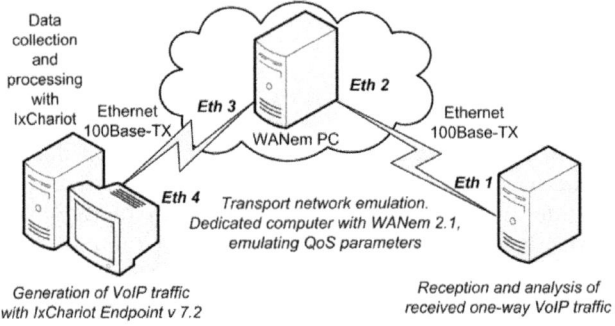

Fig. 1. Test bench for VoIP MOS evaluation with E-Model

Based on simulation results and measurements we have determined optimal shape parameter ξ giving the smallest overall MSE error of differences between measured and estimated P_{loss_wo} and P_{loss_wr} by equations (13) and (14). Optimal value of sought shape parameter ξ is around values -0.1 to -0.2 depending on actual network traffic characteristics, but in general value of -0.1 proved itself to give good results across wide range of LAN IP networks.

$$P_{loss\_wo} = \left(1 + \frac{\xi(x-\mu)}{\sigma}\right)^{-\frac{1}{\xi}} \qquad P_{loss\_wr} = \left(1 + \frac{\xi(x-\mu)}{\sigma}\right)^{-\frac{1}{\xi}} \cdot \frac{1}{2} \qquad (13), (14)$$

After substitution of equations (13) and (14) into (12), parameters $\xi = -0.1$ and $\mu = 0$ we get equation for jitter buffer packet loss without reordering (equation (15), x = buffer size in [ms]) and with reordering capability (16) where x = packet size in [ms]:

$$P_{loss\_wo} = \left(1 + \frac{-0.1x}{\sigma}\right)^{10} \qquad P_{loss\_wr} = \left(1 + \frac{-0.1x}{\sigma}\right)^{10} \cdot \frac{1}{2} \qquad (15), (16)$$

To use these calculated losses in E-Model we propose to use equation (17), which should be used in place of parameter P_{pl} (network packet loss). Equation (18) is final proposed equation for equipment impairment factor calculation in E-Model $I_{e\text{-}eff}$.

$$P_{plef} = 1 - (1 - P_{pl}) \times (1 - P_{dejitter}) = P_{pl} + P_{dejitter} - P_{pl} \times P_{dejitter} \qquad (17)$$

$$I_{e,eff} = I_e + (95 - I_e) \times P_{plef} / (P_{plef} - B_{pl}) \qquad (18)$$

Graph in Fig. 2 shows comparison of: measured MOS, MOS estimated by original implementation of ITU-T E-Model and MOS estimate using E-Model with proposed modifications in case of G.729 codec. Modified E-Model takes into account three additional parameters: network jitter, buffer size [ms] and audio packet length [ms]. Proposed modifications give more accurate MOS showing good correlation with real network conditions with jitter as opposed to too optimistic original E-Model, we received similar results for other measured codecs.

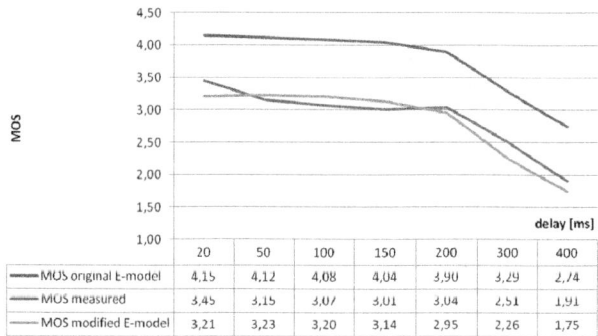

	20	50	100	150	200	300	400
MOS original E-model	4,15	4,12	4,08	4,04	3,90	3,29	2,74
MOS measured	3,45	3,15	3,07	3,01	3,04	2,51	1,91
MOS modified E-model	3,21	3,23	3,20	3,14	2,95	2,26	1,75

Fig. 2. MOS, G.729 codec, network jitter 20 ms, buffer 40 ms

4   Conclusions

We propose the E-Model improvement based on the substitution of packet loss parameter P_{pl} by the effective packet loss P_{plef} and we perceive it as the main contribution of this paper. Since the jitter J is calculated as floating average of 16 packets, we consider the characteristics constant if they are similar over recent 16 samples. When considering codec with 20 ms audio per packet, so all values are balanced after 320 ms. The transient response of real-time E-Model shows sufficiently fast recovery rate for practical purposes below 1 second.

Acknowledgement. This work was supported by IT4Innovations Centre of Excellence project, reg. no. CZ.1.05/1.1.00/02.0070 within Operational Program 'Research and Development for Innovations' and by project VEGA No. 1/0186/12 „Modeling of Multimedia Traffic Parameters in IMS Networks" conducted at Slovak University of Technology Bratislava.

References

1. International Telecommunications Union, ITU-T Recommendation G.114: One-way Transmission Time, ITU-T, Geneva (2003)
2. Koh, Y., Kiseon, K.: Loss Probability Behavior of Pareto/M/1/K Queue. Communications Letters 7(1), 39–41 (2003)
3. Mirtchev, S., Goleva, R.: Evaluation of Pareto/D/1/K Queue by Simulation. In: Information Science and Computing, International Book Series. Technical University of Sofia, Sofia (2008)
4. Dang, T.D., Sonkoly, B., Molnar, S.: Fractal Analysis and Modeling of VoIP Traffic. In: 11th International Telecommunications Network Strategy and Planning Symposium NETWORKS 2004, Vienna, pp. 123–130 (2004)
5. Morton, A., et al.: Draft new Recommendation G.1020 (formerly G.IPP): Performance Parameter Definitions for Quality of Speech and other Voiceband Applications Utilizing IP Networks, TIA (2003)
6. International Telecommunications Union, ITU-T Recommendation G.107: The E-Model, a computational model for use in transmission planning, rev. 8, ITU-T, Geneva (2000)
7. Voznak, M.: E-model Modification for Case of Cascade Codecs Arrangement. International Journal of Mathematical Models and Methods in Applied Sciences 5(8) (2011)
8. International Telecommunications Union, ITU-T G.113, Appendix I: Provisional Planning Values for the Equipment Impairment Factor Ie and Packet-Loss Robustness Factor Bpl, ITU-T Recommendation G.113, ITU-T, Geneva (1999)
9. Kovac, A., Halas, M., Orgon, M., Voznak, M.: E-model MOS Estimate Improvement through Jitter Buffer Packet Loss Modelling. Journal Advances in Electrical and Electronic Engineering 9(5) (2011)
10. Kovac, A., Halas, M., Voznak, M.: Impact of Jitter on Speech Quality Estimation in Modified E-model. In: Conference Proceedings RTT 2010, Velke Losiny, Czech Republic (2010)
11. Mirtchev, S., Statev, S.I.: Study Of Queuing Systems With A Generalized Departure Process. Serdica, Journal of Computing 2(1), 57–72 (2008)
12. Broitman, M., Matantsev, S.: Determination of Optimal Size of the De-jitter Buffer of a Receiving Router. Automatic Control and Computer Sciences 42(3), 133–137 (2008)
13. International Telecommunications Union, STUDY GROUP 12 – Delayed Contribution 98: Analysis, Measurement and Modelling of Jitter, ITU-T, Geneva (2003)

Author Index

GPSR Compliance

*The European Union's (EU) General Product Safety Regulation (GPSR)
is a set of rules that requires consumer products to be safe and our
obligations to ensure this.*

*If you have any concerns about our products, you can contact us on
ProductSafety@springernature.com*

In case Publisher is established outside the EU, the EU authorized
representative is:

Springer Nature Customer Service Center GmbH
Europaplatz 3
69115 Heidelberg, Germany

Batch number: 09478804

Printed by Printforce, the Netherlands